D0257849

pig

tales from an organic farm

pig

tales from an organic farm

HELEN BROWNING
with TIM FINNEY

First published in 2018
by WILDFIRE
an imprint of HEADLINE PUBLISHING GROUP

1

Cataloguing in Publication Data is available from the British Library

Hardback ISBN 978 1 4722 5803 8

Typeset in Adobe Garamond by CC Book Production

Printed and bound in Great Britain by
Clays Ltd, Elcograf S.p.A.

Headline's policy is to use papers that are natural, renewable and recyclable products and made from wood grown in well-managed forests and other controlled sources. The logging and manufacturing processes are expected to conform to the environmental regulations of the country of origin.

HEADLINE PUBLISHING GROUP
An Hachette UK Company
Carmelite House
50 Victoria Embankment
London EC4Y 0DZ

www.headline.co.uk
www.hachette.co.uk

To all those who care for our land and its creatures,
and especially those – past, present and future –
who support and work with the Soil Association.
Future generations will be grateful for your endeavours.

Contents

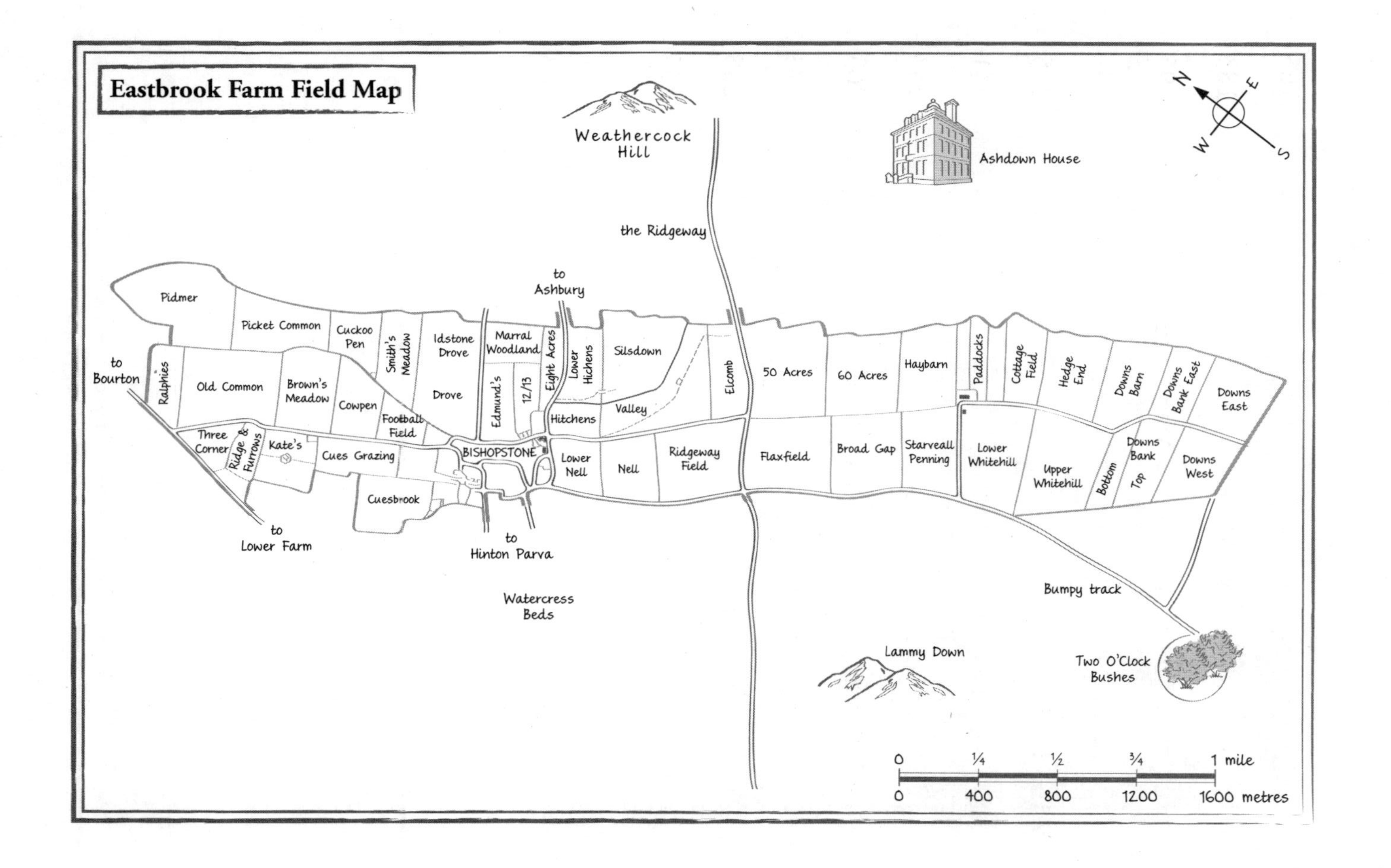
Eastbrook Farm Field Map
Weathercock Hill
Ashdown House
N
E
W
S
the Ridgeway
to Ashbury
Pidmer
Picket Common
Cuckoo Pen
Smith's Meadow
Idstone Drove
Marral Woodland
Eight Acres
Lower Hichens
Silsdown
Elcomb
50 Acres
60 Acres
Haybarn
Paddocks
Cottage Field
Hedge End
Downs Barn
Downs Bank East
Downs East
to Bourton
Ralphies
Old Common
Brown's Meadow
Cowpen
Football Field
Drove
Edmund's
12/13
Hitchens
Valley
Three Corner
Ridge & Furrows
Kate's
Cues Grazing
BISHOPSTONE
Lower Nell
Nell
Ridgeway Field
Flaxfield
Broad Gap
Starveall Penning
Lower Whitehill
Upper Whitehill
Bottom
Downs Bank
Top
Downs West
Cuesbrook
to Lower Farm
to Hinton Parva
Watercress Beds
Bumpy track
Lammy Down
Two O'Clock Bushes
0
¼
½
¾
1 mile
0
400
800
1200
1600 metres

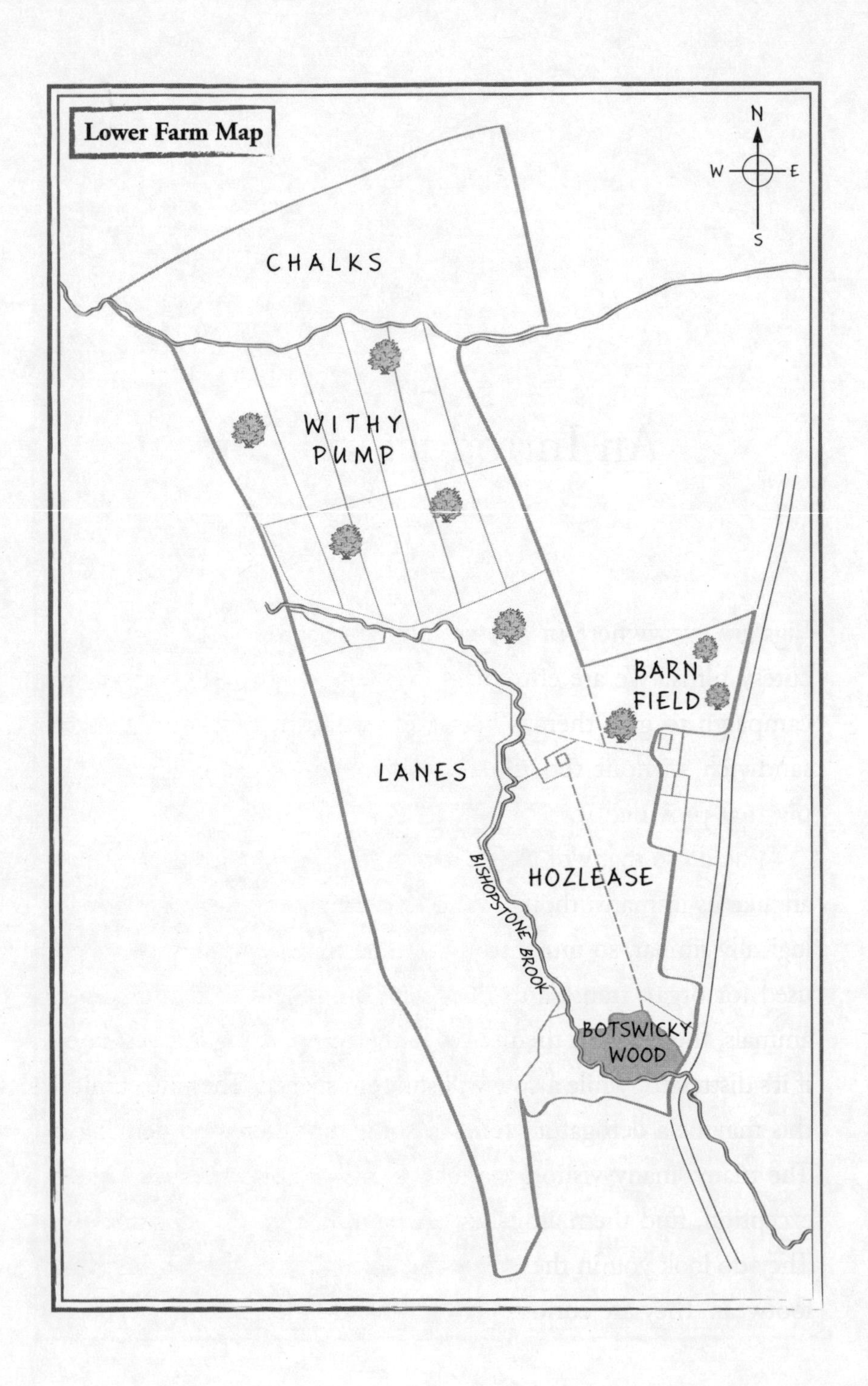
Lower Farm Map
N
W
E
S
CHALKS
WITHY PUMP
BARN FIELD
LANES
HOZLEASE
BISHOPSTONE BROOK
BOTSWICKY WOOD

An Introductory Note

Pigs are everywhere in our culture: in cartoons, in language, in cutesy films. We are entranced by them, disgusted by them, we campaign to give them a better life while tucking into a bacon sandwich without too much thought about the origins of the pig that provided it.

Why do so many of us feel drawn to pigs? Perhaps because they are like us in many, though by no means all, ways. We are physiologically similar, so much so that humans are designing pigs to be used for organ transplants. They are omnivores, and unlike prey animals, they express themselves fairly clearly. A pig lets you know if it's distressed, while a cow will suffer in silence. They are comical; this may be a derogatory term to some, but there's no denying it. The many, many visitors we take to see the pigs, almost without exception, find them amusing, enchanting and a little unnerving. They do look you in the eye, at the same time as dismantling your footwear. They are curious, inventive and independent.

They are unlike us in that they cannot sweat, so they need to wallow in water or mud when the temperature gets above 18°C or 20°C. Hence their reputation for dirtiness, I guess, but where the saying 'sweat like a pig' comes from, I have no idea; not from someone who knew anything about them. Their noses are ultra-strong and sensitive, and their sense of smell is so much more acute than ours – hence their use as truffle hunters. Given the need they have to experience the world through their noses and mouths, it seems an especial crime against pigs to prevent them from this basic requirement by housing them in barren concrete pens, or putting rings through their noses.

Humans have lived with pigs for thousands of years, notably in Asia. For the vast majority of that time, it has been a symbiotic relationship. We have fed them our waste and protected them from many of the challenges they would face in the wild, while they have provided delicious meat, well suited to preserving with salt for winter fare. Until 80 years ago, most rural households would have had a pig or two, and the autumn slaughtering, while grim from our sanitized perspective today, was a moment of community and celebration.

The breeds of pigs we have now, and worry about losing, are all a result of fairly recent human interference. The origins of many of our breeds are obscure, but most came into being in the eighteenth and nineteenth centuries, with many of the breed societies being formed only in the early twentieth century. The Essex and Wessex Saddleback societies started in the 1920s and amalgamated the breeds into one – the British Saddleback, the breed we keep here at Eastbrook – in 1967. Since then, the pig world has changed very fast, with intensive, large-scale indoor units becoming the norm, and the traditional breeds and systems very quickly became rarities.

I have been an organic farmer all my life, and pigs have been a big part of my farming career. I have never done what we are about to do now, however, which is to closely observe one particular group throughout their lives, to get to know them more intimately, and to record this faithfully. It feels quite an adventure, possibly an uncomfortable one at times. Like most farmers, I guess I protect myself emotionally with a degree of detachment from seeing the majority of our animals as individuals, though there are always some that stand out from the crowd, whilst also campaigning vigorously for a better life for pigs generally. My rural and farming upbringing has engendered a pragmatic perspective on the cycles of living and dying, and our part in that as farmers. I am aware that this matter-of-factness can be a little daunting for some, as can be our mutual sense of mischief. A degree of gallows humour is a pre-requisite for survival at times, and so far, we are survivors. I sense, though, that some huge challenges lie ahead for farmers and rural businesses.

This is not the definitive book of the pig. It is the story of a group of pigs on my farm, from birth to death. It is the story of a rural business created to allow us to keep pigs as we feel they should be kept, and some of the twists and turns of that seemingly impossible task. It is an exploration of our relationship with farmed animals, with nature, with each other, with life and death.

I write this at a moment of change, with the next generation starting to take the reins on the farm. And change in the wider world too, as the UK prepares to leave the EU, and the security blanket – or straightjacket, for it has been both – that our membership has provided for farmers over the last 40 years. What lies ahead is uncertain. This is a pivotal period, one that future generations will judge us by. There is a very short window of time to avoid the worst of the climate-change disaster that we are wreaking on ourselves, and we have already triggered the sixth mass extinction of other species. And it's not as if we are enjoying the fruits of our destruction; we are stressed, overweight and fearful.

For many years, I have been a member (and for over a decade the Chair) of the Food Ethics Council, a small charity that demands powerful thinking on how we put values centre stage in the food system. The fundamental question they ask – a question that I have asked myself repeatedly in my work as a farmer – is 'What should we do, all things considered?' I hope my book can be a useful contribution to this eternal question.

Helen Browning
September 2018

1

The Twelve Days of Christmas

21 December

It is the shortest day. The ground stiffens as dusk descends on the winter solstice. Silence, but for the hum of wings as a cloud of starlings wheels and then streams like smoke towards the hill. The year closes in upon itself: dark reflection, a suspended moment. There is still some movement in the field, however. Renegade piglets returning home after bedtime, bigger ones grabbing a last snack from the feeders before the long night ahead. A restless young sow continues to carry grass and straw to the nest she is building, even though she feels the milk heavy in her udder, and rumbling spasms starting. Finally she is satisfied with her bower, and eases herself down, cumbersome and grunting, back to the open door to block the way.

By midnight, the first piglet is born. Shocked by sudden cold and silence, she lies stunned for a moment, snuffling first breaths, then miraculous instinct prompts the scrabbling towards milk. The sow's legs are in the way, twitching as another piglet begins the pressured dive. Straw sticks to mucus, shaking limbs, heart fluttering fast, the desperate groping towards salvation. Blind persistence rewarded, the teat is grasped, so that within 20 minutes of undocking from the placenta that has sustained life until now, she is reattached to the sow. The rich first milk, the antibody-filled colostrum, flows freely.

Ten minutes later, competition arrives, and long before dawn, eight piglets are lined up on the udder, feeding and sleeping, feeding and sleeping. One last push and the placenta is expelled, steaming briefly. The sow carefully shuffles and stands; she needs to drink, and to pee. The piglets heap up together for warmth, silky little devils with their long faces and soft skin over bone. Mum is back soon, nosing them away from the bowl of her nest so that she can very carefully lie down again. This first return to the nest is a high-risk moment. The piglets are weak and huddling, so easy to tread on or squash. They squeal quietly, she grunts and totters and presents her teats again, as she will do repeatedly over the coming weeks.

Before first light, the sound of a tractor rouses pregnant sows elsewhere in the field. Breakfast is on its way. They have to wait, however, as David, the head pigman, checks the farrowing sows to see what the night has brought. Two bundles of piglets, all alive, one Saddleback litter of 13 and our Large White cross batch of eight. He brings feed, though the sows haven't got much appetite yet, and checks the piglets over while their mothers are rather apathetically eating.

He selects the two largest piglets, a boar and a gilt from the Saddleback litter, holding them close to stop them squealing, and slips them into the mound of white piglets in the adjacent arc. A strong sow can just about feed 13, but it takes its toll, both on Mum and the babies. Litters of ten or 11 should ensure that they all get plenty of milk and grow evenly. The sows seem oblivious. Unlike cows and sheep, they will accept other piglets readily if introduced soon after birth; maybe they can't count beyond five.

The sows are all British Saddlebacks, glorious black pigs with a white stripe across their shoulders. They are usually wonderful mothers, producing lots of milk if well fed, and because they carry plenty of condition (that's fat to you and me) they can feed their piglets until they are eight weeks or more, without getting thin. They are hardy, loving the outdoors, happily foraging away in most weathers. And their offspring taste fantastic. But many people are wary of too much fat on their meat, so the boars are Large Whites, a leaner breed. Crossed with a Saddleback, the piglets benefit from 'hybrid vigour' . . . a strong immune system that results from two diverse strains breeding together. The resulting piglets are predominantly white with moderate amounts of subcutaneous fat.

Every month or so, the very best sows are selected to be mated with a Saddleback boar, to produce the next generation of mothers. So amongst the hundreds of chubby white piglets roaming the field, there's a smattering of perfect little Saddlebacks. The boys will be eaten, but the girls, the gilts, as they are called, will stay on the farm to breed.

The night frost has given way to milder air, so the tractor squelches through mud as David moves up the lines of sows, checking each litter and feeding the sows who are still housed individually. The field is laid out in blocks, each holding either eight sows and their piglets, or around 80 weaned pigs, or 16 dry sows, either pregnant or with the boars. About a week before they are due to give birth, sows move into their farrowing paddocks, fresh grassy pasture subdivided by a cat's cradle of electric wire so that each sow has her own patch, with an arc – a little house – and access to water. They settle in, start rooting up the turf and making their nests. It's best to give them their own space at this stage, so that they can be fed individually by hand. If they eat too much before they farrow, they are more inclined to have problems during or after the birth. And it stops the problem of sows doing sleepovers in each other's houses. Two sows in an arc meant for one means any piglets that are born are more likely to get squashed.

Once the sows have farrowed, they need a lot more food, building up over the first ten days of their piglets' lives. The amount they need depends on the number of youngsters they are feeding; a rule of thumb is 0.4 kg for every piglet, plus 2 kg for the sow. So if a sow is feeding ten babies, she will need to eat around 6 kg of feed by the time the piglets are two weeks old, maybe more than that for the next week or two, before the piglets start eating some solid food too.

Bucket in hand, David puts the right amount of food close to the arc. Sows who have just given birth are wary of leaving their piglets and prefer to eat close enough to guard the door.

He encourages the sows out, to make sure they are shipshape, and so that he can check the litter over without being molested. Moving back and forth to the feed wagon to the rear of the tractor, stepping over the low single electric wires dividing the paddocks, he soon warms up. There's a thin veil of frost on the water troughs, which he breaks; the sows need to drink plenty of water to make all that milk.

Watching him work while I'm keeping an eye on the newborns, I reflect on how tough a job it is, out here in the winter. In a normal week, I may spend a couple of hours in the pig field, popping in to see how things are going. David is here all day, every day, in wind, rain and mud. You have to like pigs a lot to care for them outdoors in a British winter!

Once the youngest litters have been checked, David moves on to the older ones. The dividing wires have been removed in these groups, and they have one big feeder so that they can just help themselves whenever they want. Moving from pen to pen, he ensures that the feed is running freely, that the water trough is full, the fence is working, and that all the pigs are present and correct. After an hour, with all groups inspected, it's time for a break and a cup of tea.

Our sow has eaten a bit and is back with her litter, oblivious, it seems, to the cuckoos who have joined the nest. They are less than eight hours old, yet a pecking order has already been established. The firstborn – let's call her Molly; I am reluctant to allocate names as, with a few exceptions, we just don't do it, but it will make this tale easier – is especially strong, and has claimed the second teat. The Saddleback infiltrators have quickly barged their way in, bagging other prime spots. But there's plenty for all. The milk is still colostrum, rich in antibodies, which will protect the young pigs from disease. They are born almost defenceless, weighing around 0.5% of their mother's mass, with little body fat. These first feeds are absolutely vital to their survival. Over 150 g, ideally twice that, needs to be consumed in the first six hours of life, while the piglet's gut is still permeable, allowing the colostrum antibodies into their blood system. By the time the pig is a day old, the gut will have closed completely; by then, the colostrum has become milk anyway.

The tiny size of piglets compared to their mums also explains why sows have relatively few problems giving birth, unlike sheep, or cows, or us, who usually give birth to only one or two young at a time, but which are much larger offspring compared to the size of the pelvis. While David will keep an eye on sows due to farrow, and intervene on the rare occasions when things aren't progressing, it's usually best to just let them get on with it in peace.

So ends the first day of my adventure with these pigs. I came to the field before dawn knowing that the piglets born overnight would be my companions over the coming months, as I observe their lives and seek to understand them even more intimately than ever before in my 30 years of farming. It is a lottery, this choosing of a group to follow, but then, so much of life is. I am especially curious about how much they will let me into their lives, and whether getting to know them as individuals will change the way I feel about farming these remarkable creatures.

22 December

Dawn is breaking, but there's no sun. A grey, murky, morning mist is sweeping over the Downs as warmer air hits the cold land. A perfect morning for pigs, and through the mist they emerge from their shelters, trekking across the paddocks to feed, drink and undertake their morning ablutions.

Molly and her siblings seem full and contented. They are a motley bunch, I can see now, with a range of markings, quite apart from the Saddleback pair. Two, including Molly, have black heads, and two have black blotches around their middles. Even by day two, they have started to lose their rat-like look, with the beginnings of some fat appearing under their thin skins. It always astonishes me how quickly they fill out, almost as though they were being gently blown up with a bicycle pump.

Another sow in the group of eight is just finishing farrowing as I arrive. The last piglet to be born is still covered in afterbirth, but is already suckling. It's a big litter; we think we count 14 but don't want to disturb her too much. They are on the small side, and in an ideal world David would take at least a couple of them to another sow with a smaller litter, but no one else has farrowed overnight. They will have to take their chances.

It's Friday, and Christmas Day is on Monday. So along with the usual routine, David and Tony, his assistant, are trying to get well ahead before the holiday break. All of the feeders are filled up; they can hold enough to keep the pigs happy for at least five days. Every arc is bedded up with fresh straw, and as it's getting wet in the field, more straw is put by the entrance to the arcs, so that the pigs' feet are clean before they get inside. Despite our derisory allusions to pigs – usually when we are talking about each other – pigs are fastidious animals. They never soil their bedding unless they are forced to through confinement. But when the field is muddy, they need a doormat. Mud traipsed into the house will dry and turn to dust, and dust can cause respiratory problems.

The bales of straw are huge, and the guys carry a wedge from these large square bales to each farrowing arc and tease out the right amount of straw so that there are no lumps and bumps for piglets to get trapped behind. Until the sows have farrowed, they are not given too much straw. The last thing we want is a steep-sided, deep nest with piglets at the bottom. They must be able to move away from the sow easily, and she needs plenty of room to get up and down without squashing them. The first few days are the dangerous ones. When the piglets are a week old, more straw can be added, which they bury themselves in when it's cold.

While bedding-up, David checks for any damage to the arc where the wind could get in. It only takes a small hole, or a back window left ajar, for a draught to form that will chill the vulnerable piglets. However cold it is outside, the arc itself should feel snug. The back of the arc faces the prevailing wind, and with a sow inside generating plenty of heat, temperatures should stay above 10°C or 12°C.

Bedding the weaned pigs and the pregnant 'dry' sows is a cruder task. Big chunks of straw, even whole bales, are chucked into the large arcs, and the pigs have fun making their own beds.

In the afternoon, the guys start making up the paddocks for

next week's farrowings and weanings. Unlike most outdoor herds, which can stay on the same field for years, this organic herd moves across a big swath of land. Whenever a sow farrows, she does so on a fresh bit of grassland that hasn't had pigs on it for several years. When her piglets are weaned, they also go on to fresh ground. So the pig enterprise rotates around all of the suitable land – that which is flat and reasonably free draining – with new paddocks created each week, and old ones dismantled. Their dung fertilizes the soil, so that heavy crops can be grown after they leave.

To many people, farmers included, this seems like a lot of work. But David and Tony have got a slick routine, and can build a paddock in a few hours. The whole field is planned out so the three parts of the enterprise – the farrowing sows, the growing pigs and the dry sows – move along together, making best use of the space. Each paddock is 40 metres square, with two strands of electric wire around the perimeter. David and Tony start by marking the corners with strong wooden stakes, with white plastic ones holding the wire along each run. Then there's water to be plumbed in, and arcs and feeders to install. The feeders and water troughs are placed on platforms of wooden railway sleepers, which provide a solid base out of the mud for the pigs to stand on. The farrowing paddocks are subdivided further, so that each arc stands in its own sliver of ground.

It is dusk by the time they finish, so the final check and feed for new mums is by torchlight. They do the rounds together. David is on leave for the next four days, so it's good to do a handover.

It's not just David and Tony who are getting ready for Christmas. The whole farm has spent much of the week clearing muck from the dairy and youngstock, piling in the straw so that no new bedding will be needed until Boxing Day, stockpiling feed in the right places – anything to minimize work on the day when everyone wants to be with their family and friends as much as possible. As well as the pigs, Eastbrook has a milking herd of around 200 British Friesians, cared for by Teo, our wonderful Romanian herd manager, and his assistant, Andy. The cows have a calf each year, of course, and these are all reared on older dairy cows until they are weaned at around six months. So there are usually around 600 cattle, at various ages and stages, and mostly housed at this time of year, except for the dry cows and a few older beef cattle who are living on fields of turnips, with bale silage and straw to balance their diet.

We grow arable crops too: wheat, barley, spelt, beans and oats. You can understand why so many farmers choose just to grow crops. At this time of year, there's nothing much to be done on an all-arable farm. Crops don't need feeding or bedding or fencing, and they don't escape just as you are sitting down for a quiet evening, or worse still, about to eat Christmas lunch. A load of malting barley is leaving the farm today, though, so the augers are chugging away in the grain store, a sound that takes me to summertime.

There's a somewhat impromptu farm get-together this evening. Friday's always a lively night at the Royal Oak, the pub/restaurant/mini-hotel that we run in the village, with loads of locals drinking after work, including some of the farm stalwarts, and tonight there are a few more of us, though by no means all the crew. Two of the staff team are laid low with some lurgy, and there's much hoping that they will recover in time to fulfil their part of the Christmas routine. This rota has been carefully planned over the last few weeks, and the last thing we need is to have to reschedule it all. Even on Christmas Day at least four people need to be on duty: one to milk, one to look after the calves and youngstock, one to feed and bed the cows, and one on pigs.

I end up talking to David for a fair while, hearing about how he came to have a life with pigs. Raised in Fraserburgh, and still with a Scottish accent that the pigs seem to decipher more easily than do his colleagues, he started life as a painter and decorator before training at a huge pig unit nearby, never having even seen a pig before. He recounts his shock at the noise of sows in stalls waiting to be fed, and how, day one, he spilled more food than went into the troughs in his hurry to stop the screeching. He had seen some sows in straw yards, the ones that were shown to visitors, and naively asked when the ones trapped in their cages would go outside. It was a daft question. Back then, in the 80s, stalls for pregnant sows – which confine them in cages for all of their four-month pregnancy – were still legal, and indeed the norm on most indoor pig units. The sows would spend their entire pregnancy confined, only to be moved into farrowing crates a few days before they were due to give birth. His triumph, in his two and a half years there, was to ensure that all the sows had six weeks a year in the straw yards, free for a short while.

David moved between several farms over the next 15 years, before finding himself at Eastbrook. He far preferred the outdoor farms he worked on over the indoor ones, but was still surprised by our organic approach when he arrived here five years ago:

things such as moving the pigs on to clean ground, weaning much later, at eight weeks rather than the three to four weeks on other farms, the use of a traditional breed, and smaller things too, like the fact that we don't use fenders – metal barriers that attach to the doorway of the arc – to keep the piglets confined to barracks.

As we drink beer and eat pub-made sausage rolls, the place fills up. Our neighbours come in with excited children, ordering plates of chips. Hotel guests come in to dine, and the last big office party is getting underway upstairs. Suddenly, there's a holiday feel. Christmas is a bit special. Even though the farm and the pub are always working, for me it's the one time of year when emails stop arriving, when everyone does the minimum necessary, a chance to settle into the rhythm of our local village of Bishopstone and of the farm without constantly dashing off to Bristol or London. My role as Chief Executive of the Soil Association, the charity that champions a healthier way forward for farming, forestry and food in line with organic principles, keeps me away from the farm here for far too much of the time.

23 December

My daily routine now includes an early-morning visit to see how the piglets are faring. I've kept pigs for over 30 years now, but it's the first time since the very early days that I have followed a specific gang. I am fretting about the new big litter. Can they possibly all have survived their first 24 hours, so many and so small?

There are 13 piglets still with the sow. Possibly we miscounted yesterday. Tony comes over. He's found none dead this morning, but says the sows are skittish, and wonders whether there's a fox about. The worst time for foxes is usually in the spring, when vixens are feeding their cubs, and piglets make an easy meal. They will creep up and take piglets just as they are born, while the sow is still distracted by her labour, or grab a piglet that strays. It's rare to see any evidence except for lower litter numbers than we would expect, or have counted before.

The sow is number 111, and is still inside the arc feeding her young. For the first two or three days, especially with a litter this big, she will rarely leave them. Her neighbours are up and about though. Molly's mum is eating heartily, but both she and the sow with the Saddleback litter are very protective, barking at me as I come to have a look at their babies. They stand in front of their doors, on guard.

I come back in the evening. It's mild, even up here at 600 feet, warm enough to lean up against straw bales that protect the entrance to the weaned pigs' chalets and watch a large flock of lapwings wheeling and diving. They seem to have made the field south of the pig field home for now. This field, Starveall Penning, was oats in the summer, undersown with a new clover ley (temporary grass). The stubble still prickles up, even though the sheep have recently grazed it, and the lapwings seem to move between the short pasture and the empty paddocks that the pigs have finished with. Lots of bugs and earthworms, I guess. It's deeply peaceful, just the frisson of energy that emanates from the birds as they whirl and swoop, white flashing against the darkening sky.

Christmas Eve

I walk the Downs with Dog, our ageing brindle greyhound, before checking in on the sows. The farm is some five miles long, and only half a mile wide, a thin strip of land that runs from the flat heavy clay of the Vale of the White Horse to the open downland. It belongs to a large landowning institution, and my father came here as their tenant in 1950. So I have lived here all my life, and the farm is as much a part of me as my own skin.

Roughly half the land lies above the village, Bishopstone, rising on the underlying chalk to the ancient Ridgeway that traverses east to west. Wonderful dry valleys run from this ridge down to the plain, formed by meltwaters, we believe, at the end of the last ice age. Then the ground levels out for a mile or so, the chalk covered by a silty loam; only a foot of soil at best, but sticky in winter, and capable of growing good crops of grass and cereals. 'A vale farm on the hill,' my father used to say. A second rise takes you up on to the very top of the Downs, and it is here that I walk this morning.

The sealed road ends abruptly, and a track of chalk, dug many moons ago from the valley, scoured and crevassed by water running from the hill, takes you up to the top. As children we called this 'the bumpy track', and unless it's been recently graded, it still requires some careful driving. I prefer to walk, and so does Dog – the steep incline gets the blood moving. The track initially divides Eastbrook from our neighbour's farm, then veers away on to his land, up to Two o'Clock Bushes, a patch of scrub on the skyline. His downland is all in arable cropping, and there's a somewhat smelly mountain of sewage sludge near the bushes, awaiting spreading. Despite the rather strange odour, nothing like the wholesome tang of farmyard manure, it's good that this land is getting some organic material back into it. As a society, we have flushed far too many precious nutrients out to sea; recycling these to the land is vital, if we can do so without contaminating it with heavy metals or industrial pollutants.

I turn left along the high ridge. The path is springy. The soil up here is very rich in organic matter, even on the land that has been cultivated for decades now, and you can feel the sponginess under your boot. It feels wild and free, though if the wind's in the wrong direction, you can hear the hum of the M4, and see it too, carrying poor souls towards London. Before long, I am back on the grasslands of Eastbrook, and I put the lead on Dog as we slip through the hunting gate to join the pregnant ewes.

The sheep are New Zealand Romneys, bred to be hardy and easy to look after. They belong to my daughter, Sophie, and her husband, Dai, and will outwinter on the hill before lambing in late March. They are the only livestock still up here in midwinter, and once we are through them, Dog can run free. From here, the view is expansive. To the south-east there is no road or habitation in sight as the Downs roll to Lambourn some six miles away as the crow flies. Weathercock Hill runs due east, hiding the famous White Horse Hill at Uffington, a little further along the Ridgeway. Our eastern boundary is also the county boundary between Wiltshire and Oxfordshire, and a strong hedge marks it. But my eye is always drawn to Ashdown House, which nestles beyond, reminiscent of a French chateau with its perfect proportions and gold ball on the apex of its cupola, possibly the prettiest of all National Trust properties, and surrounded by woodland.

The farm stretches out below, visible to the Ridgeway. It is a patchwork of crops and grass, with the pig field, Broad Gap, pockmarked with arcs, bales of straw and tracks, providing the main action at this time of year. Seagulls are spread out across the field, crows too, and the starlings rise and fall as the mood, or the disturbance of the tractor, takes them.

I start my descent across the thick old turf of Downs Bank, through into Downs Barn where manure is spread thick on the clover ley. This is a hungry field, but by the time the soil has digested this lot, it should be satisfied for a year or two. The eponymous barn in the corner has mostly collapsed, but we leave it there to house owls. As a child I would camp overnight here, in an old caravan, to keep an eye on the ewes lambing on the Downs. Any ewes that needed care, or who were having lambs fostered on to them, were kept in little pens, 'cubs', for a few days, amongst the straw and hay bales. In between lambing checks, I would doze, watching the silent, ghostly barn owls gliding between the high wooden rafters.

Then on to Upper Whitehill, a large undulating field with a broad strip against the foot of the Downs, which is planted with a mix of sunflowers, now long-withered, fodder radish, triticale, phacelia and quinoa, to feed wild birds over the winter. There are photography hides there too, disguised as pig arcs, so that keen photographers can watch the badger sett late into the night – another of our farm ventures! Next door is Lower Whitehill. The pigs were here last winter, and now it is in turnips. Turnips are a great crop to plant after pigs, as they grow quickly, catching and holding the fertility that the pigs deposit so that it doesn't leach away during the winter; we can't afford to lose all those wonderful nutrients, and it's not great if they end up in the watercourses either. Our dry cows are strip grazing them now, getting a new section every day or two. The field is polka-dotted with lines of plastic-wrapped silage and naked straw bales; the silage provides some extra energy and variety – cows cannot live on turnips alone – and the straw provides dry pickings and somewhere less muddy to lie.

By the time I get to the pigs, it is blowing hard and spitting rain. Dog is happy to stay in the truck, and with the sows on their mettle, it's best to keep him away. Of the eight sows in the block, five have now farrowed. Two gilts have small but strong litters of seven and eight, with a hotchpotch of colourings; 111 is still womanfully feeding her 13; then there's the Saddleback litter of 11, and Molly and her siblings. There's a surprise, though. In the end section, two gilts I haven't seen here before are together in one arc with a bundle of piglets snuggled up between them. I can't count them, and I can't work out what they are doing here.

I find Tony to ask what's going on. He explains that they both farrowed unexpectedly, in with the pregnant 'dry' sows, rather than in the individual arcs we put them in a week or so before they are expected to give birth. They seemed inseparable, and in any case it was impossible to tell which piglets belonged to which mum, so he moved them into this block together. They are lying udder to udder, with the piglets surrounded by teats, and all look perfectly content. Perhaps the gilts are sisters and have maintained a strong sibling bond. It's impossible to tell without going through the records in the office.

The remaining three sows show no signs of getting on with producing piglets. One of them, a prick-eared girl who must have some 'unpure' genes in her lineage, is full of beans, prancing around in mock horror as I approach. Her bed is a lumpy mess too. I can't see her giving birth for a while yet!

Christmas Day

As far as getting some time together and unwrapping a few presents was concerned, my partner, Tim, and I had our Christmas Day on Saturday, even if our celebratory meal was a bowl of rather stodgy chilli, the result of over-enthusiastic stretching of the meat with pulses, rather than turkey and trimmings. As ever, our lives spin in different dimensions, he with the pub, rooms and Chop House (our restaurant in Swindon), and me with the farm and Soil Association. So it's a pretty cursory 'Happy Christmas' as we go our separate ways.

It's 7.30 a.m., but the dairy is already washed down and empty, the cows fed and bedded. On Christmas Day, everyone is especially keen to get started and finished as soon as possible, but Dai is still struggling with the newest batch of calves. When we foster the young calves on to their 'mums', we usually put a group of six to eight calves in the same pen as two cows. For the first few feeds, we will yoke the cows if necessary to allow the calves to suckle. In this group, however, one cow greedily adopted six of the seven calves, and for the first few days, had enough milk to keep them all happy. But the calves are getting bigger and hungrier, and the other cow needs to start pulling her weight. She's not keen, and keeps kicking the calves off as they try to drink. There's only so many times they are prepared to get kicked before they give up and get listless with hunger.

Sorting this one pen, making sure that every calf has a full feed, can take longer than all the rest of the jobs put together. Dai's patience is extraordinary, especially given that he was up at five to scrape and bed the dairy cubicles. Eventually, they have all suckled, and then Daz, one of our longest-serving farm folk, arrives with the feeder wagon, spewing a fresh line of silage mixed with grains, soya and straw down the feed passage. Although it's an 'ad lib' system, the cows all rouse themselves for the new food; it's a race to get the yummiest bits, I think. A good time to see whether there are any problems brewing, any cows that hang back who may be malingering.

On my way to the pigs, I stop to walk Dog down to the Marral woodland. This is an 18-acre wood that we planted close to the village 16 years ago. There was no woodland on the farm then, which seemed a huge shame; surely every farm should have a wood, and a hay meadow too, to provide some refuge for all the wildlife that we have displaced. Our landlords weren't too keen, however. Under the current crazy farm-support system, farmers lose their area payments when they turn cropland into woodland, which means that the land value may be reduced. In the end, they saw the merits in the plan, and Marral, which was a heavy-clay field rife with blackgrass, was planted with a whole variety of trees in different sections. Deer are a menace to new trees, so we put a high fence around it to keep them out, and created a wiggly trail with some grassy glades so that it is a perfect walk: a safe place for dogs to run free, chasing the grey squirrels, which are a great pest and have done much damage over the years.

There are no new farrowings, but I'm relieved that the 13 piglets seem active, still feeding avidly. A couple of them are lagging behind, though, with their hair standing on end and shivering slightly. The sow has cleared her breakfast, and while settling down again, has managed to trap three of them behind her. They are squawking noisily in frustration that they can't get to the milk, and given that she doesn't seem inclined to move, I lift them over to rejoin their litter mates. The squawking stops, and they muscle in to feed. There's endless jostling and play fighting in the arc. The pecking order is well established by now, but with milk at a premium for such a big litter, they keep testing their siblings for signs of weakness.

Piglets grow at an extraordinary rate. A sow's milk has twice the fat content of human or cows' milk, and over 50% more protein. Overall, her milk solids (the stuff that's not water) is around 21% rather than around 12% for cows. Human milk, however, is much higher in lactose, the milk sugar, and much lower in protein. Maybe that's why so many of us end up with a sweet tooth. Anyway, it only takes a few days for the bony little devils to turn into chubby sweethearts, with a layer of fat to protect them from the cold and provide some reserves. Once they have this insulation, not much will hurt them. This litter's not there yet; it's taking longer with so many bellies to fill.

I have to leave them, to do the Christmas Day thing with family and friends. At least David will be back later to check on them and all the newly farrowed sows.

Boxing Day

It's been a foul night. The rain and wind woke me several times, and in between I dreamed of 111 and her too-many piglets. By dawn it's calm, however, and a perfect sunrise. I start my usual Boxing Day walk, along the Ridgeway to the Idstone crossing, then south towards Ashdown House. From here, Starveall looks almost cosy, nestled down beneath the hill. It's a different perspective, not one you can imagine when you are actually on those exposed fields. I turn back though; the walk I want to do will take several hours, and I am too anxious to see how the piglets have fared. There are starlings all round 111's arc, which feels like a bad sign, and, indeed, there's unfinished feed by her door. The piglets are still feeding; this is getting ridiculous. While all the other sows and gilts in her block are eating or sleeping, piglets chubby and content, there she is, this listless sow, constantly feeding her young. She seems twitchy, which worries me further.

I go to find Tony. He is surprisingly chipper. Like me, he had heard the rain in the night and expected to be bedding-up pigs all day. But the wind was a north-easterly – cold, and blowing against the sides of the arcs – so the ground has held up well. Maybe there wasn't as much rain up here as in the village. I ask him about the sad 111, and he confirms that she hadn't eaten yesterday's food, so he got her up earlier. He too hasn't seen her when she's not been feeding her young, but thought she seemed OK when she came out of the arc. He will go back again and give her fresh food before he leaves.

The forecast is for heavy rain again tonight, starting mid-afternoon, and as the louring begins, I go back up to the field. From a distance, my heart leaps; 111 is out of her hut, rooting. By the time I get to her, she is back in again, but her feed is gone, and for the first time on one of my visits, the piglets are lying away from her, sleeping. The rain is heavy now, but I feel warmer.

27 December

It has snowed overnight. Only an inch or so, but it's a bitter morning, with a stiff north-westerly wind. Dog and I start out on what we expect to be a short walk in this weather, but having made it to the top of the Eastbrook Valley, decide we can get to the pigs on foot. Over the Ridgeway, a moment's respite from the wind, then up the hedge-line between Flaxfield and Fifty Acres, where the snow is deeper. I'm keen to see if the badger sett near the bale stacks is active; this snowy morning is a good time to see tracks.

All is pristine around the sett. Any nocturnal activity had finished by the time the white stuff fell. I know they are in there, though. The round straw bales laid out end-on have had their centres hollowed out, and there's straw around the entrance to the sett where the badgers have been using it to line their sleeping quarters. Tony mentioned on Boxing Day that he had found a dead piglet with scratch marks down its ribs between paddocks. This sett is close to the pigs, and I wonder whether the badgers have been up to mischief.

Badgers worry every farmer. They are implicated in the transmission of bovine tuberculosis (TB) from wildlife to farm animals, especially cattle, and we have had a major outbreak of the disease in the last year. Nearly 40 of our dairy cows met an untimely end because they showed an immune response when tested, and we are always worried about the pigs too. It's illegal to kill badgers, unless you are in a designated cull zone, where the aim is to reduce the population by at least 70%. This is a controversial policy, with heated debate on both sides. We have chosen to try to vaccinate our badgers, with varying degrees of success.

When the farm was surveyed three years ago, it was estimated we had between 30 and 40 badgers here. In the first year, the team managed to vaccinate just one! Year two, they got seven, and then the vaccine supply dried up, so we had to skip a year. This year they trapped and vaccinated ten, a much better result, but given that you need to achieve 70% to 80% coverage, it's still unlikely to have much impact. It's also impossible to tell whether the same badger has been vaccinated in subsequent years. Anyway, it's good to be trying to do something, and two badgers were caught at this sett; fingers crossed that they are the only two, and that they were not already infected, in which case the vaccine is useless.

The pig unit is moving this way, and within a few months will be right alongside the sett. When this has happened before, we have fed the badgers to try to stop them entering paddocks to steal feed or piglets. It might be about time to start that again.

I'm relieved to see that the water troughs are still running in the pig field. With this slushy snow there will be enough to do bedding-up, though at least both Tony and David are at work today. Then I see the latest problem. The snow has weighed down the electric fence lines, so that they are all on the ground. The pigs can just walk over them, and getting groups of pigs muddled up would be a disaster. The sows with piglets aren't a worry; they will always go back to their arcs. But dry sows and boars are another matter. If the boars get the chance, they will fight, sometimes to the death. And sorting groups of growing pigs out if they become muddled is tough, time-consuming work. We aim to ensure that the weaned pigs stay in stable family groups all their lives, as they too will fight if mixed with an unfamiliar gang. It's not so bad outside, though, as at least they can get away from each other. Indoors, there's nowhere to run.

I tie Dog up and start shaking off the fence lines, and can see Tony and David doing likewise across the field, having disconnected the batteries. All looks under control, helped by the fact that it's so cold and windy that most pigs are sensibly inside. I pop over to see my sows, especially desperate as ever to check on 111. All feed gone, that's a good sign, and the piglets are active and curious. I count them repeatedly. Still 13, I'm pretty sure. Amazing. I'm feeling a strong bond with this somewhat listless sow, making such an effort to care for her piglets, but dragged down it seems by the effort. She appears under the weather, but there is nothing wrong with her physically as far as we can tell. Her passivity reminds me of Tess of the d'Ubervilles, a favourite tragic heroine of my youth, though their predicaments are in no way similar! 'Tess's' piglets keep climbing up the sow to check me out, cheekily curious, and given that the last thing they need is a spell outside the arc on this bitter day, I leave them be.

I do a quick check of the rest of the group. The smaller litters are full of beans, and I have no worries about them at all. Then I find the prick-eared lady with the messy bed finally farrowing! Her white underbelly, swollen with milk, is towards me, and several piglets are already getting stuck in. There are more to come, I'm sure; she is enormous.

I can't stay to watch as Dog will be freezing, and it's a fair jaunt home, this time with the wind in our faces. He is pleased to get moving, and we cross Flaxfield to the Ridgeway, taking refuge along the sheltered pathway until we get to the farm/county boundary. Three fallow deer jump out of the Fifty Acres hedge, and scarper on to the neighbour's oilseed rape. Over the stile, and down the Elcombe track to the head of the valley. Wimps, we follow the valley bottom home, protected from the worst of the wind. As the ground flattens, we are joined by two red kites. They glide above us, perhaps hoping that we will make their day by flushing prey out of the snow.

28 December

A stunning morning. Some snow lingers, preserved by the sharp frost that's come in overnight. The roads are lethal, but the bright sun thaws them in parts, and provides a postcard-perfect scene.

It's a Thursday, and so David and Tony are bringing finished pigs into the Forty buildings close to the dairy. They have big bedded pens, with an outdoor run on concrete, where the feeders and drinkers are. This is home for the last few days of their lives. It's good for them to get used to concrete, which they have never seen before, so that the pens and walkways at the abattoir are not completely alien. They will go on Tuesday, and here at the Forty we have a proper race and loading ramp to make that part of the process as stress-free as possible. The boar pigs are well grown, rummaging contentedly in the straw.

A couple of families staying at the Oak are keen for a farm tour, so we load them all up into the old red Land Rover and head slightly gingerly up the hill on the slippery roads. The children are nervous at first, and hang back, which suits me fine as I want to make sure that the sows we are visiting are friendly before letting the children in with them. We check out some of the three-week-old litters, mostly hiding in their arcs, and then traipse across the frozen ground to see some older weaned groups. At about 14 to 18 weeks of age, the pigs make perfect playmates. They are bolder and curious now, and if the kids will stand still for a few minutes they will gather round to chew boots and even allow a bit of cuddling. It's a good lesson for young humans to stop trying to chase, to move slowly and carefully, so that the pigs will come over and say hello.

I take the group into one of the big metal arcs, to see how cosy it is. 'Where's the heater?' one of them asks, and is amazed that this warmth is generated by just the pigs themselves. Then off to see the boars with their harems, again careful to ensure that they are good-natured before the children get too close. No sex on view today. I often end up doing the facts-of-life spiel to younger visitors, if their parents think this is helpful!

Tony and David are busy with the water bowser. The pipes feeding water to each paddock are frozen, so water needs to be brought from the village, and each trough filled. Pigs suffer quickly if they run short of water, and a sow will drink ten or 15 litres a day to make all that milk, so it's an urgent job. But it's good to be working in clear sunshine with no wind, even if it is minus three. They have also put fresh straw around the entrance to all the arcs, after all the slushy snow of yesterday. I pay a brief visit to Tess, not wanting to hold up the families, who now have cold feet. She seems OK, but I can only count 11 piglets.

I come back later for a proper look. The piglets are sleeping contentedly, and so is their mother. They look stronger to me today – but the heap still looks too small. The last pregnant sow in this batch is nesting away, so perhaps we will have more piglets tonight.

29 December

A cold rain is washing out the lingering snow. I meet David in the pig field, and we check out the sows. There are definitely only 11 with Tess now; we search the arc and find two dead, buried in the straw. The whole litter is weak, and therefore more at risk from squashing; they huddle up for warmth, and are less able to react quickly as the sow moves around, however carefully she does so.

Even more worryingly, 'messy bed', the sow that I watched

farrowing the night before last, has only four piglets. There were at least five when I left her, and she was still popping them out. It's clear that we have a fox problem.

Losing piglets like this is always upsetting. It's such a waste of life after all the hard work that goes into breeding them in the first place. This time it affects me especially deeply. I've been watching these sows intently since the beginning of the Christmas break, and have started to become attached to them in a way that I rarely get the chance, or allow myself, to do. I'm thinking about them a great deal, even when I'm not with them, and what started as an observational experiment is turning into a much more profound engagement.

David tells me about a farm he worked on where they built a fox- and badger-proof fence around the whole farrowing unit, but it's harder to do that here when the pigs move on to fresh ground all the time. At least if we shoot the foxes now, they won't have cubs with them; I hate the thought of cubs starving because we have shot their mother.

Tess seems reluctant to get up, and she hasn't eaten breakfast. I go home for mine, and when I head back later she is still recumbent, though her feed has been eaten . . . by her, I hope. I want to see her on her feet, and check she's all right. With a bit of a struggle, she gets up, a little unsteady. She comes out of the arc and has the longest pee. It looks a bit dark, and I wonder if she is dehydrated. She starts to graze immediately, and then wanders over to drink. Now that she is up, she seems much more alert, and very keen to eat as much grass as possible. Without her in the way, a few of the piglets come out of the arc, but decide it's too cold and return very quickly. I leave her rooting away, now turning the soil with her strong nose, and hope that the fox stays away tonight.

30 December

High winds this morning: apparently Storm Dylan is on its way. I walk east in the relative protection of the Ridgeway, up to Lammy Down, the highest point around. The sky is magnificent, and the village below looks snuggled into the cleft of the valley. Bishopstone, like all the villages along this stretch of the Icknield Way, the ancient route that runs parallel to the Ridgeway through these parts, is there because of the spring line, where the chalk meets the clay. The spring that feeds the village, or did until mains water arrived, rises at the foot of the Lynchetts, terraces within the network of beautiful dry valleys above Bishopstone. It is assumed that these large steps were used for cultivation, but it's hard to imagine why. They are north-facing, and there is plenty of easier land to cultivate around here, so while they must be man-made, their origin is still a mystery. At the spring head there are abandoned watercress beds. Bishopstone was well known for watercress before the First World War, and there is still plenty there, doing its own thing in the clear, pure spring water that filters through the gravel beds, then into the stream that feeds the village pond via a series of glorious water gardens in front of the centuries-old hewn chalk-and-thatch cottages, hidden from view unless you know where you are going. Most of the farm's water still comes from this spring. It is piped to a pumping station behind the old Mill House in front of the pond and then up to a reservoir on Nell Hill and another on Lammy Down. From there, gravity feeds it around the farm and to the farmhouse. It's one of my many joys to have chalk water, unchlorinated – the most delicious drink on earth.

Blown to bits by the strong wind now, I venture off to see the pigs. Tess and her piglets are sleeping, and I can't see any sign of leftover breakfast. The gilts' smaller litters are growing fast, and they are a lot less protective now that their piglets are more robust. Molly and her gang are exuberant, playing inside the arc while their mother snores gently.

Although it's very windy, it's warmer too, and many of the older piglets around the field are out and about, playing and rooting. I trudge across to see David, who is checking the growing pigs. Their paddocks are getting muddy after all the rain this week, and they have created well-worn tracks between their houses and the feed hoppers. The wind is whipping water out of the troughs, making even more of a mess.

David confirms that all the piglets are present and correct, but like me, is still worried about Tess. She wasn't enthusiastic about breakfast, and her udder is dry. She just seems to want to eat grass, and there's not much of it about at this time of year. I have a feeling that if she had a wider range of plants to choose from, she might well be able to help herself better. This way of keeping pigs gives a far more natural environment than indoor systems, but it's still nothing like their native habitat, which would be scrub and woodland.

Henry, who is my daughter Sophie's father and still the overall farm manager, pops in for coffee, and confirms that two foxes were shot last night. They weren't right next to the pig field, but close enough, given how far they range. Let's hope they are the rogue ones.

New Year's Eve

A warmer morning. Tim has come out with me on my early perambulation to meet the pigs I have become rather obsessive about. In his dark-red trousers and walkers' waterproofs topped with an ancient pork pie hat, he stands out from the dreary midwinter landscape, and indeed from everyone else in these fields this morning. 'Fitting in' has never been part of his game plan; his maverick individuality is loved by many, and despised by a few . . . as anyone reading our Trip Advisor page will testify to! At the same time, his loyalty and exploratory nature means he is always up for whatever new adventure I suggest – just as I get dragged into his enthusiasms, drawing a line, though, at motorsports and cars in general. So my deep-dive into pig behaviour has piqued his interest. Having offered to help with the writing, it's about time he met the subjects of our investigation!

Most of the sows are out, including Tess. We've brought her some silage and Brussel sprouts that are never going to get eaten by us, as she seems so keen on Dr Green rather than her pelleted feed. All her remaining piglets look OK, but they are lagging far behind the others in the group that farrowed around the same time. Tess seems happy enough, grazing away when we arrive, but is disturbed by my offer of goodies, and returns quickly to her piglets, clearly still very protective. As soon as she arrives back in the arc, they start that familiar low squealing that all piglets make, demanding that the sow lies down to feed them. She responds with gentle grunting: 'I'm coming, I'm coming.' They latch on, and I'm sorry that I have disturbed her break from these avaricious youngsters.

Molly's mum and her neighbour are much calmer now. They snuffle at the proffered silage, though they are still clearing up breakfast. Both groups of piglets are growing fast, and this morning they come out of the arc for a bit, not for the first time, but it's good to see them getting their noses into the earth. The soil is their protector, just as it is ours. Their first contact with it allows them to ingest the microflora, all the millions of bugs that we know so little about, which helps their gut stay healthy. As most of the antibiotics that are given to pigs are to stop diarrhoea – what we call 'scours' – keeping their gut microbiome in good shape is vitally important. My friend Annabel who farms pigs in Jamaica, always puts soil into the pens at the first sign of scours, though I don't know of any indoor farmers who do that in this country. It's maybe been just too easy to chuck the antibiotics in. Soil also gives them iron, another reason to get their noses into it as early as possible. Indoor piglets will be injected with iron, but we have never needed to do that here.

At this time of year, the piglets won't spend very long outside of the arc until they are a couple of weeks old, whereas in the summer, they are often out and about within a day or two. They are much like us, preferring to snuggle up in the warm in the winter, but once they have a good layer of fat, they will spend a few hours a day outside when it's mild enough, running around, playing and rooting.

I mentioned that unlike most outdoor sow herds, we don't use fenders, unless we are fostering piglets. These are metal frames that fit on to the arcs, and stop the piglets moving away from the shelter. It makes it easier for the pigman to catch the piglets, in order to cut their teeth and tails, and to wean them. Given that we don't need to do these mutilations – indeed, organic rules forbid it – and that we want the piglets to get up close and personal with soil as soon as possible, we only use a fender in exceptional circumstances, and then just for a day or two. David was surprised at this when he came here, but soon saw the benefits. An added one, which I hadn't cottoned on to until he mentioned it, is that the bedding in the arc stays much drier. The piglets mostly pee and poo outside with no barrier in their way, so there's much less condensation in the arc, a build-up of which can lead to respiratory problems, especially in winter when the ventilation flaps are kept closed to stop draughts. This means David has to bed up less often – an added bonus.

In their first couple of weeks, the piglets are not usually handled at all. No iron injections, no teeth clipping, no tail docking. Because they live outside, and will do all their lives, they will not suffer the stress that causes so many pigs to become aggressive with each other. A pig's tail, dangling away, is such a temptation for a bored pen-mate. Of course, they shouldn't be dangling. A curled-up tail is the sign of a healthy, contented pig; a dangling tail says, 'I'm a bit (possibly very) off colour.' Anyway, dangling or not, once blood is drawn, it can soon be bedlam, very unpleasant indeed. Hence the fact that most pigs who are destined to live indoors for most of their lives (even if they are born outside) will have their tails cut short to try to stop this cannibalism.

It occurs to me, writing this, that I've spent most of my time talking about the problem animals. We have around 1,500 pigs and over 500 cattle at various ages and stages, plus 300 breeding ewes. But it's the few vulnerable ones, the premature calf, or an oversized litter like Tess's, that take all the time and worry.

The investment in getting the basics of good husbandry right, like feeding the mother well during pregnancy, ensuring that vaccinations are done, providing plenty of space and exercise, making sure the piglets get plenty of colostrum immediately after birth, all mean that the work and expense is less, and that the animal will live a healthy, productive life. In my role with the Soil Association, we spend much time promoting healthy school meals, getting kids out on to farms to engage with real life, not just a computer game, and trying to persuade the government that investing in our children's health will, as with our animals, pay dividends in the longer term. I still despair though. As with many farmers, we give more attention to the balanced diet and to the environment that our animals live in than society sometimes gives to rearing the next generation of humans. 'Prevention is better than cure' is an old adage, but one that I want the government to act on, and now is a good time to be making that case, with all our farming and food policies in flux as a result of our decision to leave the EU. This must be a good moment to rethink how we could invest in happier, healthier people and animals. In my role as Soil Association Chief Executive, that's certainly what I will be pushing for over the coming months.

I'm back in the pig field at dusk for a final check around, and to say goodbye to the year. There's no setting sun, but a full moon is rising above the Downs. 2017 has not been the easiest year, but it's another one gone, and given the too-short time we have on this earth to marvel and to learn, each one must be celebrated, revelled in. I just hope the revelries tonight won't disturb the pigs. There will be fireworks in the village for sure, and the cattle can be startled. Up here, though, over a mile away, the animals should sleep in peace.

New Year's Day

A grey, mild, still morning to start the new year and my last chance for a while to walk miles across the Downs and spend time with the pigs on a weekday. I do my favourite ramble, high on the ridge. Little has changed since I walked this way a week ago; at this suspended moment of the year nothing grows, and even the earthworms work slowly. The light is always different, however, and the birdlife creates animation above the stillness of the land.

The pigs, of course, change daily. It has been engrossing to observe this batch of sows as they have farrowed, to see the piglets evolve from skinny scraps of life into robust, chunky little puppies. Molly and her siblings have more than doubled their birthweight in the last 12 days.

Tess is out and about again, and seems better every day. I

would like to see more milk in her udder though. She looks like a dry sow, but she must be making some milk or her 11 piglets would not survive, let alone grow. And they are growing, albeit slowly compared to the others in this group. I watch her for a while once she goes back into the arc. Her tiny piglets are hopping in and out of the shelter, on to the straw pad around the entrance, and occasionally, courageously, out into the field. A quick explore, then they scamper back. They really are little darlings!

This gang of sows and gilts have all got quite used to me over the last 12 days. The gilts especially, once so wary, now come over to say hello, and chew my boots if I let them. They are fine with me going to check out their babies, while a week ago they would rush back to stand guard. The two gilts who are living together are especially charming. Their combined litters are all sorts of colours, including some brown ones with Saddleback-like white stripes. One of the gilts, 302, has flecks of brown in her black coat, and on her tail too. There must be some Duroc genes in there somewhere. We had Duroc boars here for many years. They are a hardy, tasty breed with ginger coats, originating in the USA, though the parent stock of the original Durocs probably came from the UK. They produce robust darker-coloured piglets, again without too much fat – although all our pigs will have more fat than the standardized ultra-lean white things that are the staple of the pig industry today.

The abattoirs hated the coloured pigs, however, as the bristle is hard to remove, and customers aren't used to a dark rind on their bacon. In the end, we switched to the Large White as our 'terminal sire', so that the piglets are mostly pale. This means that they are rather more prone to sunburn – that's certainly not a problem at this time of year – and the boars are a bit less hardy too. The boars need to be brought on to the farm when they are fairly young, and ideally in the summer, so that they can grow on more slowly, and get used to their new free-range life before the weather gets cold.

Occasionally, if we don't have enough Saddleback gilts, we will keep a few cross-bred ones that would otherwise have gone for meat. They will make perfectly good mothers, even if they don't match the rest of the herd. So this brown-tinged gilt must be the granddaughter, maybe the great-granddaughter, of one of these, and her multicoloured piglets are a throwback to the time of the Durocs.

I reflect on the stark differences in the experience of these piglets and their dams during these early precarious days in midwinter, compared to their indoor brethren. The sows have been able to nest, farrow and care for their piglets without the constraints of the farrowing crate, still used by nearly all indoor systems, though there is some experimentation with less constraining 'free farrowing' approaches on some progressive farms. They are able to choose whether to be inside a reasonably snug arc, or to range outside to feed and root. I think for the sows, this system wins hands down.

For the piglets, it seems to me that there's a weighing-up of all the freedom they have and the avoidance of mutilations, like tail docking, with the additional risk they run from their unconstrained mothers, careful though they are, and from predation. Out of the 62 piglets that we know were born to the seven sows and gilts, 58 survive – just over eight per mother, which is not a great average litter size. We would usually expect them to rear ten or so each. In addition to this, one sow, Mrs Messy Bed, as I've come to think of her, lost nearly all her litter to a fox – we don't know how many because we didn't have a chance to count them. The four that died were found in the arcs, presumed squashed. How do you decide which is best, to have freedom, with its attendant risks, or a safer life that gives much less opportunity for exploration and fun? Maybe it's not unlike the dilemmas we constantly face as parents; our instinct is to protect, yet at the same time, we know that risks must be taken to make the most of life.

These first 12 days are the riskiest for the young piglets, and life should become rather more straightforward from here on as they grow and develop. I hope so anyway. I have expended much emotional energy on these scraps of cuteness, and I'm heavily invested in their survival. At the same time, the debates we are having about the future direction of the farm, and the far-reaching decisions that will be made about the future of all farming by our political leaders, will need my full attention over the coming months. Sitting on the straw pad outside Molly's arc, and watching the swirl of starlings in the darkening sky as the piglets climb over me, I'm deeply aware of how much is at stake.

2

To Meat or Not to Meat

3–7 January

My personal 12 days of Christmas at the farm are over, and it's time to re-engage with the outside world. Before I do, though, I steal a few moments to see how the pigs are getting on. It's a blustery, wet morning, but many of the older pigs are out around the field, busily rooting up the moist soil and trekking across their paddocks to eat and drink. The weaned paddocks have big bales of straw outside to allow the pigs to make a deep pad to hang out on. Unless it's pouring with rain, there will usually be a group lounging around, enjoying the fresh air, and possibly the view.

Molly and her cousins are all abed, though, still reluctant to spend much time outside in this inclement weather. The sows pop out to feed and rootle, but don't stay away from their youngsters for long. It's their large bodies that keep the arcs so warm, and they seem to know that they are needed as a radiator, as well as for their rich milk. I duck inside with one of the gilts, 307; she has become very relaxed with me, coming up for a scratch whenever I am around, and seems totally chilled about sharing her arc for a bit. Her piglets are soon boot-chewing. But I can't stay long. The world of people and politics awaits.

I have been Chief Executive of the Soil Association for seven years now, and it is amazing to be able to help a charity that is dealing with the most important things in life, how we eat, and how we produce our food. These are things it's easy to take for granted, even though the evidence of how wrong we've been getting it is all around us. We have squeezed out most other species, have degraded much of the planet's best soils – and thus our chances of feeding ourselves well in the future – treated our farm animals poorly, and fed ourselves poorly too, so that very many of us are malnourished, not getting the nutrients we need, while at the same time we are increasingly obese through eating too many empty calories from sugar, refined carbohydrates and poor-quality fats.

The Soil Association has just celebrated its 70th year. For much of this time, our message has been a candle in the wind, warning of the crazy short-term thinking that has led to all these problems, as well as to accelerating climate change, nutrient wastage on an epic scale, and the antibiotics crisis. People have disrupted the natural world for millennia, but our increasing numbers, and the power we can exert through our technologies, has allowed us to accelerate this impact extraordinarily over the last 100 years. This is a time for wisdom and long-term thinking; both have been in very short supply.

The damage we are doing to the world and to ourselves is becoming so apparent that even the most sceptical are at last starting to accept that we must do things differently. Then, Brexit. Unwelcome as it is to some of us, it creates a moment where everything is up for grabs, for good or ill. We have the opportunity to show a better way forward, one that combines traditional knowledge with the right kinds of innovation, to improve our soils, put animal welfare centre stage, farm without pesticides, and feed people well. Or we could end up with low standards for animals and the environment, imports of poor-quality food, and see farmers – especially the smaller ones – going out of business.

The new year always starts with the Oxford farming conferences, where many of these debates will be brought to life very vividly. There are two of these now, one the very long-running 'establishment' event, and the other, a renegade upstart, the 'real' farming conference. The first is predominantly men in suits, the second better gender-balanced, and lots of woolly jumpers. They run consecutively and exemplify the way the world is changing. The challenger event is quickly oversubscribed, with over 1,000 delegates this year, and a huge waiting list. It has a tremendous energy with lots of sessions running in parallel on a huge diversity of topics. But the mainstream Oxford Farming Conference is changing too, with sessions this year on farming for biodiversity, the digital revolution, and even a smattering of female speakers, heaven forbid.

Thursday is the big day at Oxford, when the politicians come to set out their stalls. Michael Gove, the Secretary of State for Environment, Food and Rural Affairs (Defra), was a surprise, and not universally welcomed, appointment six months ago. Defra has a hugely enhanced workload as a result of Brexit, as they are responsible for the majority of the regulatory changes that leaving the EU will entail. It's not just Gove's appointment that has been a surprise, but the way he has thrown himself wholeheartedly into the role, engaging energetically with folk from all quarters, and setting out some progressive thinking about what the future should look like for farming and the countryside.

He is speaking at both conferences this morning; luckily they are only 500 metres apart. First, it's the older Oxford conference, with 400 farmers, business people and advisers in the Examination Schools; he is appearing alongside Ted McKinney, from the USA Department of Agriculture. Unusually for Mr Gove, it's a prepared speech, with some new angles. As well as the lines we have heard before, an emphasis on environment, animal welfare, the importance of soils, and aiming to be the best, not necessarily the cheapest – all music to my ears – he mentions a new food policy, a gold standard for food products, and the connection between farming and our health. This last point is the cornerstone of the Soil Association's philosophy. As our founder, Eve Balfour, put it: 'The health of soil, plants, animals and man is one and indivisible.' What a breakthrough it would be to have policies put in place that would take this core organic principle and make it a reality for everyone.

But, of course, the big question is what sort of trade deal will we reach with the EU, and with other nations? Will we be forced into World Trade Organization standards, so that we have to take food produced to lower standards than ours? Farmers are rightfully nervous that they will be undercut on price, and the public are largely outraged at the idea of chlorine-washed chickens and hormone-treated beef. Ted fuels these fears, in my head at least, with talk of rolling back regulation, and veiled threats about the use of phyto-sanitary measures as barriers to trade.

I dash off to the Real Farming Conference in time to hear Gove's second engagement with farmers this morning, and this time it's straight into questions. After the strait-laced, respectful audience an hour before, this must be some contrast. Oxford's glorious town hall is full to bursting, not just on the floor but five deep in the balconies too. The atmosphere is electric, energy fizzing. Concerns about the future of small farmers and the opportunities for new entrants are countered by Gove, who seems confident that land prices will fall once subsidies go and tax breaks are removed and that this will solve the problem. He suggests that more funding might be diverted into organic research, and that help might continue to be given to those who want to convert their farms to organic methods – but no commitments are given to those who already are organic, despite the well-proven environmental benefits.

One question from the audience goes to the heart of the world's dilemma: how can we maintain the GDP growth that all economies rely on to sustain living standards, without destroying nature and the prospects for future generations? The minister doesn't really answer this, but he at least acknowledges that it is a very valid question.

*

This evening it is the Oxford Union Debate. The motion is: 'By 2100, meat-eating will be a thing of the past', with George

Monbiot, activist author and columnist, as the leading protagonist for the motion. He takes a gentle, genial approach with an audience he has no chance of winning over, while still laying out the facts with devastating clarity: the livestock industry is responsible for 15% of greenhouse gas emissions, more than all forms of transport combined, and with a growing population, is a massively inefficient way of feeding ourselves to boot. He quotes research that suggests we could feed a population of 200 million in the UK if we switched to a plant-based diet. The farmer who opposes him responds with some of the lines I've used myself in the past: there are many places where crop growing is impossible, and only animals can make good use of the terrain; that farming animals sustains many rural communities, particularly in the uplands. Contributions from the floor are mostly in humorous vein, all very boysy, several fart jokes at the thought of all those vegetables. I feel I should say something, especially as speeches for the motion are hard to elicit, but this ribald dynamic is not one I will play into well, and I hold my tongue. Indeed, no women speak. It is Guy Singh Watson, founder of Riverford, the pioneering and hugely successful organic vegetable delivery service, who eventually cuts through with a short emotion-filled comment: 'This is serious. Many people will die if we don't sort climate change, and the meat industry is a major contributor.' His intervention breaks the jollity, and brings an awkward pause before the seconders begin their statements. Phil Lymbery, CEO of Compassion in World Farming, who has written extensively on livestock farming's impact on biodiversity, is all smooth rhetoric and clear facts: we use 80% of our land to produce 26% of dietary protein. All seem to agree that we must at least reduce meat-eating, and that less but better – pasture- not grain-fed, humanely reared – is the way forward.

It seems remarkable to me that no one defends the status quo: the meat industry as it exists today. Even in this bastion of the farming establishment, the consensus is that we must reduce our meat habit, and take great care with how we produce what we do eat. It's a big shift in a few years. I recall being upbraided by the executive of the Meat and Livestock Commission (an industry body funded by levies from farmers and processors) when I was a Commissioner 15 years ago, for daring to say this publicly. Times and views are changing fast.

I stay behind to chat with the speakers, keeping an eye on the 'aye' and 'nay' doors, to see how many vote for the motion. I'm surprised by the numbers going through the 'aye' door, and indeed, the count reveals that 120 went out that way, compared to 276 who rejected the motion. For a mostly farmer audience, that's an extraordinary result.

I spend Friday at the Real Conference, mostly in the farming sessions as I'm reporting on them at the end-of-conference plenary. A good chance to listen to some fascinating speakers talking about the discoveries they are making. Many of them are part of the Duchy Future Farming 'Innovative Farmers' Network that the Soil Association has helped establish, and it is brilliant to see how the support and small pots of funding have led to some surprising new approaches.

I'm also intrigued by the advances in 'small smart machines', aka robots, and how they may help us get off the tractor treadmill.

Up to 90% of the energy going into tilling the soil is dealing with the problems caused by heavy machinery. It's a circular problem. The soil gets compacted by tractors, so it requires bigger tractors to pull bigger machines to repair the damage. It would be amazing to have little robots running around planting and weeding; they could go on to the land when a tractor can't, and could navigate their way around the trees and hedges that often get ripped out, or not planted in the first place because they are a nuisance for big machines. I'm trying to envisage how they could help with the pigs. We do a fair bit of damage to tracks around the pig field in winter as we take feed and bedding around. I'm not sure that robots are going to help us with this for a while.

*

It's only been three days, but I'm desperate to see how the piglets are faring when I return home. Early on Saturday I am back in the pig field, nervously checking on Tess. It's a huge relief to find all 11 piglets still alive, and they have grown a bit too. It's Tony's weekend on duty, and as he puts out the feed, all the sows come out and get stuck in. Tess seems to have rediscovered her appetite, and although they are still so tiny, her piglets nibble at her breakfast too. They look much more vigorous, and I'm reassured by the amount of scampering and playing they are doing; it's always a good sign when they have the energy for fun.

Molly and her siblings look great. They are growing quickly, and must weigh at least 3 kg now, well over twice their birthweight. They don't bother with Mum's breakfast, presumably still satisfied with milk. The two inseparable gilts are on great form, their multicoloured piglets bold and curious, chasing under the wires that divide the sows and chewing my laces. They are starting to become distinguishable by more than their markings. The brownest of them is an especially cheeky chap, always the first to come up to me, then playing scared when I move, bounding away, zigging and zagging.

Tony tells me that a huge dog fox, 'as big as an Alsatian', was shot last night, with half a large piglet in his mouth. Finally, we may have got the main culprit, caught in the act. Litter numbers have been low for a couple of weeks, and there have been some mysterious disappearances of older piglets too. He has found a mauled dead piglet this morning, and I discover another as I walk the paddocks. Hopefully there will be no more.

It's been a busy week on the farm, catching up after the holiday period. The usual stuff, mucking out, repairing machines (there's always something that needs mechanical attention), mending feeding troughs and fences. But we also grabbed some time on Tuesday for a planning meeting with consultants who are helping us think through the complexities of bringing in the next generation, my daughter, Sophie, and her husband, Dai, while ensuring that the farm can provide for all of us, long-term.

Sophie is now 27, a tall, strapping, beautiful lass who bears witness to the fact that the benign neglect of her mother has done mercifully little damage. She has her faults, of course, as even I will admit: she makes me look tidy by comparison with the physical chaos she leaves in her wake – and yet she somehow manages to organize her life and workload impeccably, miraculously flying through exams, and covering more ground in a day than most do in a week. Even as a baby, she exuded calm wisdom, and from her earliest years, you could ask her any question and get a thoughtful and often thought-provoking answer. After doing her first degree, in Biological Sciences, she worked for three years as a knowledge transfer associate with the Food Animal Initiative near Oxford, and then decided that she just had to do her veterinary degree, if she could get a place. She did, at Nottingham University, where she is now in her fourth year, and so although she had already started living with Dai, she has been away from him and the farm for fair chunks of the last four years.

The two of them met through Young Farmers while he was managing a herd of cows ten miles from here. Dai spent his childhood on a Welsh smallholding, and so the land is bred into him, much as it is with Sophie. Energy and ambition flood from his pores, and now that his rugby days are over, that rugged energy is mostly diverted into farming. He has worked here for the last five years, building a flock of sheep on our and our neighbours' land, and taking increasing responsibility for the wider management of the farm too. When we acquired Lower Farm, the 200-acre block of land that adjoins Eastbrook, they moved into the cottage there, where they live with their five dogs (four of them for working the sheep) and two horses – rather lovely creatures that never get ridden as the two of them are much too busy for fun.

Having been given such an opportunity to start my farming career at a young age, I am keen to do the same for Sophie and Dai. Far too many farmers hold on to the reins for far too long, driving the bright ones to seek their fortunes elsewhere, or at least excluding them from decision-making until late in the day. But sorting the transfer of responsibilities is not straightforward, as we are finding, especially as Sophie is still not on the farm full-time. They are very keen to expand the number of cows we milk, and so we are applying for planning for a second dairy and winter cow housing at Starveall, at the southern end of the farm, on free-draining land that is perfect for the spring calving herd they want to establish there. But this is also the area that is most suited for pigs and the arable crops that follow them. In the passing of the baton, the pig herd that I have spent the last 30 years developing, and the sausages, bacon and pork that we sell successfully around the country, are under threat. While overall, I think their plan is a good one, and, indeed, it makes sense to produce milk and meat from grass rather than having to buy in the grain that the pigs consume, I'm still grappling with the financial risks – this is such a big investment – as well as the more emotional feelings that I have about scaling back the enterprise that I have put so much of my life into over the last 30 years.

*

The conversation with our advisers today is helpful. Having had a chance to look at our current finances and future projections, including a second dairy, they suggest that it may make sense to maintain at least some pigs, to spread the risk and improve overall profitability. That's music to my ears, of course, even though I know there are many flaws with the pig enterprise. With Dai and Sophie's energy and enterprise, I feel that we could make some progress in improving the system further. I've been away too much over the last ten years to drive new approaches forward here.

It's such a balancing act, trying to work out how to spend my time and attention. I love the farm hugely, and yet have chosen to work away from it much of the time, so that I am always just dashing around the place catching up at the weekends. There's rarely the opportunity to develop new thinking, or even when there is, no chance to follow through on it properly. Luckily, we have a great team here, and I don't have to worry much on a day-to-day basis; I know that everything will be kept up together. But we haven't developed as much as we might have over the last decade.

I weigh up the impact that I can have. To have a farm that illustrates what good organic farming looks like, and to be pioneering new approaches, is a valuable contribution. Then again, the chance to influence the direction of all of food and farming, to lead an organization right at the heart of both the debate and the practice, should lead to more widespread change than can flow from Eastbrook alone. Therefore I somewhat neglect my businesses here, and hope that I am indeed doing something useful in the big discussions that are raging about how we will feed ourselves in the future. At the same time, my life as a livestock farmer, with such a dependency on meat and milk for our living, puts me in a good position to understand the huge challenges that farmers face in moving to a 'less but better' world.

If the Oxford debate is anything to go by, there seems to be little dispute now that we who eat too much meat need to eat less, so that those who eat too little can have their fair share. If we all eat as much meat as the average American does – some 275 g/day per person – in the future, then we will certainly struggle to feed everyone adequately, without destroying what is left of nature, and doing massive damage to our environment too. But what does that mean for farmers like me? Animals are a key part of our fertility-building cycle, making good use of the clovers that build the nitrogen in the soil, so that we can grow crops without energy-expensive manufactured fertilizer. And while water efficiency is one of the much cited downsides to eating beef rather than soya (for instance), does this matter in places where there is usually plenty of rain, and no need to irrigate? While there is no doubt in my mind that we should reduce our consumption of animal products, it still seems to make sense to keep ruminant livestock in areas of the country and the world where there are great grass-growing conditions – though none whatsoever to be keeping them in areas that rely on irrigation.

However, if we are going to eat less meat and dairy – and current thinking is that our health and that of the environment would be in a safer place if we limited our consumption of meat to around 70 g/day per person – we need to be eating more of something else, more fruit and vegetables, more nuts and pulses, things that we grow too little of in the UK. There is a host of challenges to increasing the production of these, as I know only too well from the experimenting we are doing at Eastbrook. It's a big transition to make, one that will require research and investment in processing infrastructure and markets, and for tree crops especially, financial support for farmers during the time lag between planting and any commercial return.

Back with the pigs by late afternoon, still pondering these thorny debates, I notice how some of the sows have rooted up all the pasture in their section, while others have hardly made a mark. Then I see that one of the gilts has slipped under or through her fence; having demolished her patch so that it looks like a ploughed field, she is making advances on our listless Tess's pristine terrain. Avidly grazing and digging, squelching through the wet soil, she is in no mind to go back, and I have no boards to shepherd her. I'm a bit worried that she may barge into Tess's arc, and try to stay there for the night. In the end I have to leave her to it, in the hope and expectation that she will return to her own piglets once she has finished excavating her neighbour's patch.

3

It Began Like This

My life with pigs started late. My father did not keep them here; cows, beef, sheep, but no pigs. Like many people, I'd hardly seen one until I started my degree in Agricultural Technology at Harper Adams in Shropshire. We were lectured on them by 'Piggy Sadler', who was definitely a man of his time. There was only one way to keep pigs in his eyes, and he proudly took us to the 'state-of-the-art' farms nearby. As a naive farmer's daughter, coming from a farm which, while cutting edge in many ways (not all of them good), kept all its animals at pasture during the clement months, I was shocked to my core. Yes, I had stumbled across *Animal Machines*, Ruth Harrison's landmark book that exposed the inhumanity we were inflicting on our farmed livestock, in my school library many years before. But seeing it in the flesh, the sow stalls, the farrowing crates, the weaner decks and sweat boxes, and to have this presented as 'the way', was deeply disturbing.

Most people will wrestle with the issue of taking life so that we can eat. I remember my father, hardly a sentimental man, who spent much of his spare time shooting, fishing and hunting, saying, 'Who am I to play God today?' as we selected lambs for the butcher in the wooden pens behind Starveall buildings. You cannot escape the responsibility, as farmer or consumer, and I have often said that those of us who cannot kill should not eat meat, or dairy – or even eggs, as the male chicks are despatched as useless soon after hatching. Yet death is not the problem, much as we have a problem with it. As a farm child, we encountered death every day. My brother and I earned our pocket money ferreting for rabbits, which we skinned and sold. My father would give us sixpence for every rat's tail we brought home; we used terriers to catch them in the farm buildings. And if an animal was in irrevocable pain, a lamb born disabled, a rabbit with myxomatosis, a partridge wounded by a gun, it was our duty to kill it as quickly and humanely as possible. I hated it, but I did it.

We are squeamish about death, as we spend our lives contemplating our own. But death itself is not a welfare issue, though the manner of it, done badly, clearly can be. For many animals, like the ones I saw in the 1980s pig purgatories, death would be far preferable to the lives they led. My resolve became clear: to try to develop and support a good life for farm animals, for as long as people choose to eat meat.

Piggy Sadler gave us a project, the theoretical establishment of a 400-sow breeding unit. Starting as I meant to go on, mine was an outdoor system, with sows taking some of their diet from grass and other forage crops. I received a D minus, and a public dressing-down: 'Pigs don't eat grass!' It was just the challenge I needed to prove him wrong, and to spend a lifetime promoting a better way of keeping pigs to as wide an audience as possible.

My practical life with pigs started soon after taking on the management of the farm in 1986. Concerns about animal welfare were just one part of the angst I had developed about farming. It was becoming increasingly clear to me that we were also squeezing out nature. Farming has always done this, of course, but with ever more powerful machinery, sprays and fertilizers, we were, and still are, wreaking intolerable destruction on all other species. I had grown up with the orthodox assumption that being a good farmer (that is, a good person) meant extracting maximum yield, whether from an acre of land or an animal. But the cracks were apparent; even at Eastbrook I could see that the wildlife was vanishing as hedges were pulled out to make way for ever bigger tractors and more chemicals were used year on year. I started to become interested in organic farming as an alternative approach.

As you can imagine, this was another unacceptable subject at Harper Adams in those days. In three years of study, there was only one lecture that mentioned it, and while the physics and chemistry of soil were well taught, its biology barely featured. When an extraordinary opportunity emerged for me to spend my 'industrial placement' year as the research assistant on the first Ministry of Agriculture-sponsored project on the comparison between organic and 'conventional' farming, the college almost refused to let me take it. I lobbied hard, and in the end, they relented.

That year was the best possible preparation for my future. I was working for the ADAS Soil Science Department, then still a government agency, based in Bristol, part of a knowledgeable team and under the mentorship of Roger Unwin, who had a genuine interest in organic methods. Most of my time was spent at Rushall Farm, owned by Barry Wookey, who had converted this large landholding to organic production in the early 1970s, partly due to his concern about the demise of the grey partridge. The farm is on the edge of Salisbury Plain, around Upavon, and the trials involved matching and comparing the Rushall organic fields with those of similar soil types on neighbouring farms.

I monitored everything, from earthworm populations (extracting them with formaldehyde in square quadrants, then hours in the lab identifying them all) to weed and disease levels in the crops, the amount of phosphate and potassium in the soils and the quantity of nitrate that was leaching through the soil profile (with a scary-sounding 'neutron probe', which required me to carry a radioactive warning on my car). I learned more in that year about the things that matter in farming than I had done in all my studying time. How nutrients cycle, and how to stop their loss from the soil; the lives of pests and weeds, and how to prevent damage through interrupting their life cycle; the value of manures, and the way they feed soil microorganisms; how to construct a fertility-building rotation.

By the time I came back to run the farm a year later, I knew that I wanted to try my hand at farming organically here. From what I had seen, the system had lots of potential, which perhaps wasn't being fully realized on the farms I had seen, and which certainly wasn't appreciated by either the farming community or policy makers. While I wasn't convinced that it was the answer to all the world's problems, it did seem to offer solutions to many of the things I cared about – plummeting biodiversity, animal welfare and climate change especially. For someone looking for a lifelong challenge, something to get my teeth stuck into, developing organic farming at Eastbrook felt like just the thing. In many ways, I saw it as a continuation of my research; a chance to keep learning.

I had plenty of other challenges in those early days. I was young, 24, and all the staff I was now managing were men, many of them twice my age. And they were extremely sceptical, both about me, and about these organic interests that they caught wind of. It was an incredibly stressful time.

One field was already organic, as I had persuaded Dad to convert it so that I could continue my minor obsession with phosphate release from calcareous soils, the subject of my final degree project. I decided to run it as a trial site, and at the same time see if this 20-acre field, then known as Dennis Hunt's after the farmer who had this land before us, could sustain a livelihood. This was another interest of mine: how much land did someone need to make a living?

My great friend from Harper Adams, Kate Hobsley, was a London girl, energetic, practical and sparky, with no obvious way into farming, despite her enthusiasm. She joined me at the farm, and together we developed a huge range of small-scale enterprises on the field. We planted various mixtures of clover, ran wheat variety trials, experimented with inter-cropping (growing two or more crops together), dug raised beds for vegetables, and planted an orchard. Helped by the farm handyman, we built two hen houses, and filled them with free-range laying birds. And we bought our first two pigs.

*

The choice of Saddlebacks was as much chance as design. We saw some advertised not too far away, at Heytesbury, near Salisbury, so we went to have a look. Gilbert Ryall was the breeder, well known for his pedigree stock. He was a character, as they say, and somewhat surprised, I felt, by having these two young and clearly inexperienced potential pig keepers turning up on his doorstep. Under his expert guidance we bought two gilts, both in pig, which we named Gloria and Ethel for no good reason that I can remember.

There was, still is, a barn on the roadside edge of Dennis Hunt's field, then part-filled with straw. The pigs started their life there, with an outdoor area to explore as well, and even when we supplied them with proper arcs, they always preferred to sleep in the barn.

A couple of months later they farrowed, producing the beginnings of the Eastbrook pig dynasty. Between their and our naivety, it didn't go very well, and I recall that only five piglets were reared successfully from that first attempt. Still, it was a start. We kept the females for breeding, and somehow got the sows back in pig again; I think this required a return trip to Gilbert for stud services.

Our field was soon transformed from a rather boring grass ley into a hive of activity. The farm staff, while still convinced that this was all madness, were at least impressed by the yields of forage we produced, especially from the red clover trials, all without artificial fertilizer. Farmers, seduced by the sales patter of the agro-chemical companies, had forgotten how productive leguminous plants like the clovers can be. Indeed, one of my motivations in trying to farm organically was to get rid of the stream of reps who were constantly at my door, making the case for one product or another. I became increasingly determined to see how much food we could produce without all these inputs, a bit like an athlete deciding to compete without the drugs.

I quickly realized that by diversifying and intensifying the range of crops, especially vegetables, we could grow much more food from this 20 acres than the field had yielded before. You can grow a lot of vegetables on 20 acres; the problem is an economic one, the cost of the labour to sow, weed and pick. All this complexity takes more thought and management too. Much easier just to grow one thing, and have a few big machines that plant, spray or harvest in next to no time. The final problem is how do you sell it? One large-scale commodity, and it's easy, even if you might not get the price you need. Lots of small-volume crops, plus eggs, and soon meat too, and it's quite a headache.

Kate and I started to sell our produce from a stall on the side of the road by the barn. We probably got a bit of a reputation as the mad lasses who were flagging down cars, demanding that their occupants buy something. Then we would be running around the field picking whatever it was that they wanted, things that we had run out of or not thought to put on the stall. It was hardly an efficient use of our time. So to make our lives easier, or so we thought, we bought a shop.

Meanwhile, the pigs were well entrenched. Their next litters were a better effort, born in the barn rather than their arcs, and as we still hadn't got the hang of electric fencing, they mostly ranged where they felt like it. I have a vivid memory that has stayed with me always, and which was the inspiration for our brand image many years later. One of the sows, a couple of fields from home, walking back to her barn in the dusk with her litter following behind. These were pigs living their own lives, in their own way, accepting our care, and especially our food, but largely independent of us. That vision, of animals who, even though domesticated, can have some real autonomy, some control over how they live their lives, some chance to experience the highs and lows of what life has to offer, just as we would wish to, has shaped my practical farming life immensely.

*

Our pigs and our marketing ventures have always been inextricably linked. Although we started a farm shop in the local village of Shrivenham in 1987, selling much of the produce from our field, vegetables and eggs as well as meat, it soon became clear that it was the meat that was enabling the fledgling business to pay its way. We needed to take at least £2,000 a week to pay for staff and all the overheads, and that's an awful lot of vegetables! Meat is the most expensive part of our diet, and it was as a butcher's shop that we could make the business work. We sold our own beef and lamb, as well as pork, and even started to rear some chickens for meat too, setting up our own small slaughtering and dressing facilities for these. All the rest of the stock had to be slaughtered off-farm, and returned to the shop as carcasses.

Within a year or so of starting our trials, I had enough confidence to begin to convert more of the farm to organic methods. We had two dairies at that time, one at each end of the village. I decided to convert one of them and to compare the performance against the other. As the silage for the cows was grown on fields away from the dairies themselves, we started to convert other fields too, as they came into the ley phase of the rotation.

By now we had ten breeding sows, and they moved on to some of this newly organic land. But we could only increase the numbers in line with the amount of pork we could sell. Unlike beef and lamb, where the costs of producing them organically are not very much more, pigs eat a lot of cereals and protein crops, and these cost almost twice as much to buy as the chemically grown versions. Although we were producing most of their feed on the farm then, we still needed to price it into the pig enterprise at the value we could achieve if we sold it off the farm. So the higher price of feed, plus the fact that pigs who run around having fun need more of it to gain weight, meant that we needed to charge a good deal more for the meat. There was virtually no market for organic meat back then, so we needed to create our own.

One or two supermarkets were starting to get interested in selling organic food, however, notably the now-demised Safeway. I had always been sceptical about selling to supermarkets, afraid of being yet another casualty of the huge disparity in power between them and farmers – but when an intermediary got in touch and said that Safeway wanted to stock organic pork, and could we supply it, I said yes. I wanted to turn my small herd into a more substantial enterprise, one that would allow me to employ a full-time, experienced pig person, rather than asking the team to look after them on top of their day jobs.

So we expanded, first to 20 and then to 50 sows. This may not sound a lot, but 50 sows produce around 1,000 piglets a year, and our little shop sold at best 300. We were due to start supplying Safeway in May 1989. Then, just three weeks before the planned launch, we were told that they wanted 100 pigs per week in order to start, and we only had 20, so the deal was off.

This was disastrous news. These pigs had cost much to produce, and now we had no market for them. If we sold them as non-organic, we would lose a huge amount of money. Of course, this is what we had to do for a bit, while we tried to work out a solution.

We were already making our own bacon and ham in the shop, curing it in brine, the traditional Wiltshire way. Once it is cured, meat keeps much longer, so we started to cure as much as we could. We knew someone who was selling into posh London clubs and restaurants, and he started to tout our pork and hams too. The sales started to build, but the shop was far too small for all this processing. We began to convert the stable block behind the farmhouse into a butchery, to cut, cure and pack the pigs.

Our London friend was proving unreliable, so we bought our own van, and started taking orders directly. We also had the brainwave of starting a mail-order business; another example of us doing things a decade too soon. But it worked, after a fashion, and we were soon despatching polystyrene cool boxes of meat by courier all over the country. Now there were two butchers, a delivery driver and a superwoman (not me!) taking orders, pitching sales, organizing deliveries. And that was just the start.

Back on the farm, Daz had taken over as pigman. I'm never quite sure how this happened. Daz was at the farm when I came back, but a youngster, unlike most of the rest of the crew. The previous farm manager, the talented James Phillips, had given this Mohican-sporting young tearaway a chance, and he had turned into a supercharged sprayer operator, addicted to his fancy MB Trac (green mean machine of the 80s) and possibly some less-legal substances too. His badge of honour was that he could spray 600 acres in 24 hours, helped no doubt by the less-legal substances. There are a million stories to tell about Daz, but suffice it to say for now that clearly his chosen career was under threat, as the sprayed area reduced year on year, and he somehow switched flash tractors for muddy pigs. So much for my 'experienced pig person', but he cared for them enormously, and did a great job given the level of our knowledge at the time.

The wider world was starting to take an interest in this madness. Troops of government officials and journalists stopped by, and in 1990, heavily pregnant with Sophie, I was awarded 'Farming Woman of the Year'; there were very few to choose from at that time. We were of especial interest because it was a large-scale farm, and I was a tenant with a rent to pay. To date, organic farming had been seen by many people as either a rich man's hobby, or a hippie dream. Here was a farm that was paying its way, with no outside sources of income to prop it up. Furthermore, we were doing something else that farmers had forgotten to do for decades, connecting with the consumer; in fact, even remembering that there is a consumer, someone who ultimately needs to be happy with the quality of the food we produce. Since the advent of deficiency payments and marketing boards (which bought all you could produce at pre-set prices) and then the Common Agricultural Policy (CAP), farmers had been able to just decide what to farm, and get on with it, without ever thinking about whether anyone wanted to buy it, or how to sell it.

Our attempts to sell our wares more directly were not straightforward, however. The shop was going quite well, and for a while we had a second one in Newbury, within a larger store. The mail-order side of things was growing too. And we were building up quite a following for our meats, mainly in hotels, restaurants, delis and independent retailers. This soon became the biggest turnover part of the venture, and our little processing shed was running flat out. There were constant battles with Environmental Health: did we really need to build a tunnel from the van to the door, so that carcasses were not exposed for a second to that ultimate pollutant, fresh air?

Despite all this activity, it was hard to get the business as a whole to pay its way. It didn't help that it had become fairly complicated, and we were struggling to understand how the different bits were faring financially. We needed an experienced management accountant to provide the information that would help us make the right decisions about what to focus on, and what to ditch. But like many fledgling businesses, we struggled to justify the substantial cost that this would entail on such an insubstantial turnover. Big mistake. If I'm ever asked to advise a start-up these days, my first word of warning is to invest in understanding your numbers from day one. It wasn't that we didn't try, after a bit anyway, but we had our fair share of duffers who generated endless spreadsheets and no analysis.

Having outgrown the butchery at the back of the farmhouse, and possibly tried the patience of our Environmental Health Officer just that bit too much, we started to look for bigger premises. We ended up in Witney, some 25 miles from the farm, sharing a larger processing plant with an existing business. We had started to sell fresh beef, pork and lamb into Waitrose, under our Eastbrook Farms Organic Meat brand. Pret A Manger, already a high-street success, decided to see if organic would work for them, and developed, of all things, an organic lamb sandwich with us. I spent a couple of days training their food prep teams on how to make them; talk about the blind leading the blind.

With all these new chunks of business, we were running into a serious case of carcass imbalance, the plague of the meat industry. You can sell five times more loins say, but no one wants the legs. Which is a very big problem. Enter Tim Finney, the answer to all our prayers for a seasoned professional with a background in food to lead us to salvation!

I had known Tim for a few years. He was the editor of the farming and environment programmes for BBC radio, so was often interviewing me on some subject or other. With a long interest in food, he had decided to stop talking about it, and try his hand in the real world, and having looked around for opportunities to get involved with a progressive food business, he landed on us. In truth, the talents that he had – a sharp mind and sometimes sharper tongue; a deep interest in politics, including the politics of food and farming; and an encyclopaedic knowledge of sport – were not really what we needed. But he had, and still has, huge energy, optimism and drive, and these are things you can never have too much of, even if they need a bit of leavening with commercial realities. Together we have somehow weathered many storms, from foot-and-mouth disease, to closing down bits of our ventures that were failing, to starting new initiatives that we felt might have better prospects. Although he had little practical experience in marketing and business leadership – the very things he came here to do – our combined enthusiasms and naivety have drawn us into multiple escapades, and the development of a nest of businesses that somehow work, most of the time. Our division of responsibilities over the years has been quite straightforward: I look after animals when they are alive, and he takes over when they are dead. The secret of a successful partnership!

*

Today our pigs reach customers far and wide, through supermarkets like Sainsbury's, Ocado and Morrisons, box schemes (such as Riverford), and we even export to Germany, Denmark and Italy. We no longer make the products ourselves, but work with partners who are much better at this than we ever were. In many ways, our greatest pleasure is feeding people directly, through our village pub, the Royal Oak, run by Tim and now with 12 lovely rooms all themed after fields on the farm, and through the Chop House restaurant that we started a couple of years ago in Swindon. Very few brands genuinely come from a specific farm, and many customers come to the farm each year to meet the pigs in person. Tim has more time with the live pigs than the dead ones these days, taking groups of visitors to see them in our old red Land Rover.

When Tim arrived on the scene, he was determined to focus our branding, and indeed was responsible for making my name synonymous with pigs in many people's eyes. He was clear that it should be my name on the label, 'a real person, and a woman to boot', rather than a farm that nobody's heard of. And then came the orange pig that I love so much with its silly party hat, and mysterious blue landscape with hills and trees, born of my encounter with our first sow and her piglets, ambling home across the fields at dusk.

As well as being a genuine farm brand, it's also rare for a herd of pigs to be bred for decades on the same farm. The modern way is to 'depopulate' as disease levels build up, culling all the sows and starting again with 'clean' stock. Our way has been to allow the herd to evolve on the farm, to enable generation-to-generation adaptations to the particulars of the environment here. New genetics need to come in, of course, through the boar lines we use, but the maternal lines all track back to those two Heytesbury gilts, and they are part of the DNA of our farm too.

When I explain this to Molly later on she seems not the least bit interested, but then I was never that keen on family history either when I was a child.

4

A Week with the Soil Association

8–14 January

I make a pre-dawn visit to the pigs before getting on the road to Bristol for my first full week in the office since before the Christmas break. My gang can barely be bothered to acknowledge me. The most I get is a sleepy grunt. But everyone seems to be in the right place, which is a bit of a relief after the evening before when I had to leave the escapee gilt to hopefully make her way home to her young. She is back in her arc, and Tess lies disturbed only by her hungry piglets. They are trying to feed when I arrive, while every other litter is still dead to the world.

It is light, but only just, when I get to the Soil Association HQ in central Bristol. Hopefully an hour in the truck has evaporated piggy smells as I have a full day of meetings ahead. My Soil Association life complements my farming life very well in many ways, even if it can all feel a bit of a stretch at times. I think it's helpful to be wrestling with the realities of getting a farm and food business to work, and taking that practical experience into our campaigning and influencing work. On a Sunday, I may be checking the pigs, or counting earthworms or reviewing the performance of the dairy herd, and then on Monday I may be meeting with Defra officials to advise on soils policy, or speaking about animal welfare to a public audience. It's a wonderfully varied life.

Over the past 70 years, the Soil Association has been making the case for a farming and food system that tackles the global challenges of climate change, the massive biodiversity crash and human malnutrition, and to ensure that we tackle these challenges in ways that are fair and equitable, both to people and the animals we have responsibility for. This is, of course, a very shared endeavour; so many brilliant people and organizations are working tirelessly for change, and working better together at this pivotal time, in some quarters at least, than is sometimes the case. So it's important to focus the Soil Association's efforts on the areas where we have special expertise and solutions to share, or which are in danger of being neglected by others.

With this in mind, we have been making the case for soils and trees, both vitally important in preventing climate change, and also helping us adapt to it. Animal welfare is a key area for us too, of course, and, bringing all this together, organic farming as the best, well-proven way to ensure that consumers can have confidence that the food and other products they buy will enable their values, and their health, to be better protected. The right approach to research and innovation is therefore crucial;

so often it is driven by the need of companies to keep selling things, rather than by what the world actually needs – which is often better ways of doing things, rather than more products. The Soil Association has been putting farmers in the driving seat of their own research, supporting them with grants through the Prince of Wales's Charitable Foundation, and the idea is catching on.

It's when I have the chance to help the team on the issues where I have personal insights that I feel most useful; for example, my experience with establishing our agro-forestry system and running such a variety of farming enterprises at Eastbrook has given me some perspectives that might not be quite as vivid if I had just read about them.

These practical insights may be of even more value in the months and years ahead. With all the decisions that need to be made as a result of the UK leaving the EU, on trade relationships, on labour issues, on the support – if any – that farmers will receive in the future, this is the best opportunity for decades to create long-term change.

It's not enough to just grow our crops and rear our animals in a gentler manner, but crucial to help people eat well too – both for their health and for the environment – especially those who are on lower incomes, who may not have the choices that I am lucky enough to enjoy. So we focus on working with schools, hospitals, care homes and businesses to put healthier food on their menus, and encouraging government and local authorities to emphasize quality rather than just price for food in public settings. Our Food for Life programme has been a huge success, with over half of all primary schools in England now serving food to our standards – but there's loads more to do. We want all children to be learning about food and where it comes from as a core part of the curriculum, visiting farms and growing fruit and vegetables in schoolyards. We tend to work in the most deprived communities, commissioned by local authorities, and often supported by trusts and businesses too, to help tackle the malnutrition that undermines children's health and educational prospects.

Part of the transformation required is structural: we need to design our food and farming environment so that it is inherently healthy, rather than fundamentally sick. High streets in the poorest areas of our country are full of fried chicken and betting shops; how can people be encouraged to take better care of themselves and their children if they are surrounded by the wrong kinds of temptation? This is just one small example of the major shifts we must make, and which we can show are perfectly possible with a bit of effort and investment.

My working life, then, is spent trying to support and inspire this change, and this can take many different forms in the course of a week.

Today, for instance, I have a fascinating chat with a journalist about the potential for trees to provide protein for livestock feed. This is an area of great interest to me, given that much of the protein fed to our pigs is soya, usually produced far from the UK. Although it is organic, it's harder to be completely confident about the integrity of the supply when it's coming halfway around the world, and it would be far better if we could produce all the feed we need close to or on our farms. Some farmers are experimenting with growing soya in the UK, with newer varieties that are more tolerant of our cooler conditions. There's interest in insects too, as a protein source. And one day soon, we may be able to start using meat and bone meal, or other waste foods, for pigs again; after all, pigs have traditionally been a way of recycling waste, and I would much prefer to feed them that way, rather than have them eating food that could be feeding us more directly. However, the use of animal proteins for pigs and poultry was banned after BSE, even though they are omnivores like us. Swill feeding was stopped after the 2001 foot-and-mouth disease crisis, because it seems likely that poorly prepared swill was the root cause of the outbreak. So pigs and chickens, who need considerable amounts of feed to grow as fast as they do, are eating grains and protein crops that could be used for human consumption. Even though they are efficient converters of feed into meat, compared with ruminants, it's a far from perfect situation.

Trees are another opportunity to replace soya and grain in the diet. We all know about acorn-fed pigs, but we don't have the oak forests to do that here. Nuts could be another option. At Eastbrook, we are planting a whole range of trees, including oak and various nuts, to be able to experiment with these ideas. Trees take many years to mature, though, so this is a long-term game, and very speculative at the moment. I am delighted that journalists want to write about these issues; the more people are interested in and support these new approaches, the more chance we have of government taking notice.

*

I often don't get out of the office all day, which is tough for someone who is happiest in fresh air, so it's a pleasure to have a brisk walk across town this afternoon, to meet with a new alliance of people and organizations who recognize that reversing the degradation of our soils, here and globally, is the most important thing we must do if future generations are to be able to feed themselves. Soils lock up more carbon than all above-ground vegetation combined, and could hold a lot more if farmed correctly. Given the knife edge we are on, with only a few short years to prevent the worst ravages of man-made climate change, protecting the carbon stores in soils, especially peat soils, and promoting techniques that will help them store more, is a no-brainer.

The message, as we agree today, seems to have got through to government, and to most farmers too. At a recent event in the House of Commons, there was not one dissenting voice in the packed-to-the-rafters room. The challenge now is to turn the rhetoric and good intentions into practical action.

Back at the Soil Association HQ, the last meeting today is on our next appeal to members. We do a couple of these each year, on issues where we need the resources to do more work. The team have asked me if I will write a compelling letter to our supporters on animal welfare; they know it's a subject very close to my heart, and one that hangs in the balance due to our withdrawal from the EU.

I've already described the outrage I experienced when I first saw the so-called 'state-of-the-art' pig and poultry enterprises in the mid-1980s. Things have improved a fair bit since then: sow stalls have been banned, and laying hens have 'enriched' cages, with a bit more space. There's still much I would like to change, of course. Our view is that all farm animals should have the chance to range freely outside when the weather is good, and the opportunity to fulfil their instincts; for a pig, that means rooting, nesting and playing, for starters.

Government ministers say that animal welfare standards will not fall as we leave the EU. But then they voted against including the vitally important recognition that animals are sentient beings in the Withdrawal Bill. We, along with many others, were vocal in our disappointment; this principle, like the polluter pays and the precautionary principles, underpins much of our legislation and commitment to high standards.

The government has acknowledged that animal welfare should be seen as a public good, and so be eligible for financial support under whatever renationalized farm policy is finally agreed. That's great news, but we need to help shape the detail of what this may mean in practice, a big chunk of work for the team.

The biggest threat of all is what kind of trade deals we may do once we leave the EU. There has already been an outcry about the idea of chlorinated chicken coming from the USA. The issue is not so much the chlorination itself – in the UK bagged leafy salads are chlorine-washed, for instance, and no one seems too bothered – but the sense that this is used to compensate for the poor hygiene and welfare in farms and slaughterhouses. I'm even more worried about cheap, feedlot-reared and hormone-treated beef, which could undercut our farmers' largely pasture-fed beef. US pig farms are monstrosities compared with ours too. Could all the advances we have made here be undermined by imported food that is nowhere near the standards we want and expect?

I spend the evening trying to convey the urgency of all this in the letter to our members and supporters, in the hope that they will want to help.

*

These dark winter mornings are so frustrating. It's pitch-black when I arrive at the field on Wednesday, keen to say hello to the pigs and to see how they are faring before the start of the working day. I stumble across the rutted tracks, with only the flashlight on my phone to stop me tripping over the fences, but despite my efforts, I get little reward. The families are all sound asleep, and not at all interested in rousing themselves for their pre-dawn visitor. I am muddy now, and cold, and they are warm, snug and still snoring. Pigs are so much more sensible than people!

Later that morning, I head to the station to catch a train to London, where I'm due to have a quick lunch and catch-up with my old friend and ex-boss Fiona Reynolds.

Fiona was Director General of the National Trust for over ten years, and for a short part of that time, I was her Director of External Affairs. It was a hugely enjoyable and stretching time for me, and amongst many other things, helped me understand more fully the mental health benefits of spending time in nature and in places of beauty. I'd always wondered how so many people can live in cities, often without a patch of green or a tree in sight. I know that I get withdrawal symptoms after 24 hours away from the countryside! My team at the Trust was doing a review of the science of this, and indeed, the evidence is clear. We are healthier and happier when we have regular contact with nature, especially woods. Just like pigs and other farm animals, our move away from a life in-tune with the natural world to a largely urban existence has happened in what is no more than an evolutionary blink of an eye. Just like pigs, most of us have become incarcerated in a concrete jungle, but our genes haven't changed that much. However domesticated we think we have become, the increase in mental health disorders may be at least partly to do with our rapid divorce from nature.

It's a very brief catch-up, as we both need to get to the first meeting of the Royal Society of Arts Commission on the future of Food, Farming and the Countryside, an initiative that I helped establish. In these complex times, it seems vital to draw all sides of the debate together to try to find solutions that will be both bold and pragmatic, and ensure that we think long-term about the way food and farming will support our health, and the health of nature too.

This first meeting develops into a fascinating, wide-ranging debate about the issues we might choose to concentrate on. We discuss the proposed work programme and the solutions-focused approach the team have suggested will work best. There's been endless analysis done by many reputable organizations on the problems; what is needed is a more creative and inclusive brainstorm of new or undeveloped answers. We go around the table taking soundings on what our top three issues might be for starters. There are some great insights alongside our wish lists: 'How would you make things happen by default (rather than through constant policy meddling)?' is one of them that sticks with me.

An issue I raise is whether lab-manufactured meat, and indoor/urban farms, such as those being developed for fish, salads, algae and novel fungi, may reduce our dependence on land for farming. Today, land is the battleground, an increasingly impoverished battleground as we destroy the fertile soils that have developed over millennia. We are constantly expanding the land used for farming, often chopping down rainforest or ploughing out native grasslands in the name of feeding the growing population. Often, we are just feeding the animals that feed us. If we could satisfy our carnivorous instincts through a meat-like substitute, this could free up vast areas for nature. It's not an idea that plays well with most livestock farmers, however, and I understand their fear. I love my pigs, and am not sure that I want to bring about a time when they won't be part of our farming life. But the world is changing very fast, and when we remember the many things – such as the internet and smart phones – which would have felt like science fiction 20 or 30 years ago, it's not inconceivable that before long much of our food may be generated from novel raw materials in high-rise, energy- and nutrient-efficient factories. The way we farm now, all our trials and tribulations with animals, weather and crops, could soon be a thing of the past.

The next day, I have an especially early start to get to the London Wetland Centre in Barnes, to hear the Prime Minister launch the long-awaited 25-year Environment Plan.

The media headlines are largely about plastics, following the dramatic rise in public awareness of this huge issue since David Attenborough's *Blue Planet* series screened. Even for many of us deeply involved in green thinking, it's no bad thing to have this centre stage; after all, we are all culpable. Weaning ourselves off plastic is a massive challenge, and seems to me to require a fundamental redesign of our food system, where so much of the single-use plastic is to be found. Even our organic bacon and sausages are encased in plastic, to extend shelf life and protect the product in the long supply chains that most of us have been sucked into. A more direct path from field to plate must be a big part of the solution, surely, so that people can re-use containers, and buy fresher food that would need fewer preservatives and less storage, as well as less packaging. Another challenge to factor into the re-engineering of food and farming systems.

There's plenty of brilliant stuff in the Plan, but little commitment on timings, funding or the legislation that will be required to deliver some of this. It's a good start, but these issues will need to stay in the public's eyeline if they are to be more than pretty words.

I'm heading back to London again the next morning, this time for Defra's Productivity Group. This is a gang of industry leaders convened by the department to check out some of their thinking. Today's meeting is rather taken over with updates on all the moving parts that are in play at the moment, including the Agriculture Bill, the 'command paper' that will precede it, the Industrial Strategy, and of course some debrief from yesterday's Environment Plan launch. This does all take some keeping up with, with the pieces of the jigsaw and their latest timescales constantly in flux.

'Productivity' and the UK's failings to achieve a sensible level of it, is a regular lament in the media. There seems to be little attempt, however, to explore what this actually means. Is it about overall output, or efficiency ratios, such as the amount of labour it takes to produce a given output? It seems to me that we are often rather schizophrenic about this. On the one hand, we support anything that increases employment; in the next breath, we want more production per labour unit, which usually means replacing people with machines.

The castigation about 'productivity' hasn't escaped farming, hence, I guess, the existence of this group. For most farmers, productivity is usually interpreted as yield, so that a farm that grows more tonnes per acre, or rears more piglets per sow, is seen as more productive. Not for the first time, I raise the issue of definition. Officials seem increasingly clear that it is the efficiency of input use that they are interested in, especially inputs of non-renewables or potentially damaging inputs, like pesticides and fertilizers. The last buzzword for this was 'sustainable intensification', a phrase designed to please both environmentalists and hard-core farm businesses, but which landed uneasily with both.

This terminology might seem like a rather academic debate, but it can drive the culture of farming. Farmers, like any other community, want to be respected and admired by their peers. I grew up with the aspirations of all young farmers at the time, to grow four tonnes of wheat per acre, and get my cows to produce 10,000 litres of milk each year. Since the Second World War, yield was what mattered, what was praised and encouraged by governments, and then by the EU, until we had butter and grain mountains that cost a fortune to deal with. The environmental, social or animal welfare costs were entirely ignored, except of course by those tedious greenies.

As every business guru reiterates, 'measure what matters', but so often that translates into 'what's measured, matters'. Yield is easy to measure, easy to boast about. The most cited performance measure for pigs is how many piglets are weaned per sow per year. A 'good' performance in intensive units might be 24; a really 'excellent' farm is getting close to 30. This is achieved by selective breeding over many years to increase litter size, even though that means that many of the piglets will be small and therefore more vulnerable. So the rationale for confining the sow in a farrowing crate increases, as small piglets are more likely to be crushed by Mum, especially as all this breeding focus on fecundity, and other metrics like rapid growth and the efficiency of converting grain into meat, means that few farmers have concentrated on mothering aptitude. Then it's about reducing the amount of time that piglets spend with their mother, as she won't become pregnant again until she stops lactating. Piglets are weaned at between three and four weeks old; in the USA, always at the forefront of animal exploitation, this is apparently reduced to two weeks on some farms. This in turn means that the piglets lose the protection of Mum's milk while their gut is still immature. They will usually go into 'flat decks', cages with mesh floors, kept nice and warm, with easily digestible creep feed regularly laced with antibiotics or other antimicrobials like zinc oxide, to keep the inevitable scours (diarrhoea to you and me) at bay.

I took a group of Danish pig farmers to see my pigs recently.

The Danes, of course, are renowned as expert pork producers, and have 'productivity' figures that are the envy of UK farmers. Their challenge, they openly admitted, was mortality. Around 30% of their piglets born will die before weaning. They were astounded that our mortality usually runs at under 10%, around the same as the quoted figures for indoor farrowing crate systems in the UK. But as our sows will usually have 10 to 12 piglets, rather than 16 to 18, or more, and only farrow at most twice in a year, rather than two and a half times, the Danes will still wean many more pigs per sow each year. If the measure is the kilos of pig weaned per sow each year, then we win hands down. Our piglets will be three times the weight of their indoor counterparts by the time they are weaned at eight weeks or so. The moral of the tale is that maybe that 'productivity', however it's interpreted, is only one measure of many that we should be evaluating, and even then it needs looking at from every angle. And all the things we care about need their measures, and to be assessed together, not in separate silos. I'm sure my pigs would agree!

Looking back over the week, which is not an untypical one, and only fragments of which I've captured in this diary, makes me realize what a range of subjects I end up dealing with. It can feel a bit overwhelming, but at the same time, it's fascinating to be involved with so many pieces that make up the jigsaw of the better future we are striving for. I can't know all that I need to on every subject, so at times I reassure myself that the most important contribution to make is to bring organic values to every conversation. If we are guided by these values, of care, health, ecology and fairness in all that we do, we won't go too far astray.

5

Survival Tactics

At last it's Saturday, and I can reconvene with my pigs. I haven't seen them since last weekend, and I expect that there will be some changes. As I approach, I notice that the dividing fences between the sows have been removed, and a large feeder on a platform of wooden railway sleepers installed in the paddock. We do this when all the sows in the group have farrowed, and the piglets are at least ten days old. The sows can then feed as and when they want to, and the feed itself is better protected from rain and birds. The mums will eat what they need, depending on how much milk they are producing, and it won't be long before the piglets follow suit. I can already see some of them nosing around the base of the feeder, nibbling on pellets that have spilled from the hopper. The big-bottomed ginger pig is jumping on and off the platform, seemingly just for the fun of it.

Having been separated from each other for a few weeks, the sows are still re-establishing their pecking order. Pigs are very hierarchical creatures, and will scrap to establish who is boss if they are with unfamiliar animals. The odd squabble here and there is underway, but it's all very much handbags at dawn. Messy Bed has me in stitches. She is as aloof as ever, wafting around at some distance from the other sows, who are snarling at each other rather half-heartedly, then occasionally she wheels around, and makes a rapid charge at the group, looking momentarily quite ferocious, before resuming her Madonna act. The piglets keep clear, very wisely.

Molly and her gang look great. They are three weeks old, and have almost doubled in weight this week. There's a lot of coming and going between the litters, mock battles and romping, before rushing back to Mum for reassurance.

There is no sign of Tess, though. She's in her arc, and I'm shocked by the state of her piglets. They haven't grown at all; indeed, they have gone backwards. Most of their spines are visible, their tails are droopy and they are covered in long coarse hair rather than the velvet fluff of their bubbly neighbours. Classic signs of malnutrition. This sow is not producing enough, if any, milk.

Perhaps in desperation, the youngsters come to see me. I go to the feeder and take a double handful of the small pellets back to the arc. Scattering the feed in the straw, I'm amazed to see the piglets immediately snuffling around for it. Backwards and forwards I go to the feeder, carrying the pellets by hand as there's no bucket in sight, and then Tess rouses herself and starts to feed too. I wonder whether, in her listless state, she has neglected to eat since doorstep deliveries stopped a few days ago when the group was amalgamated and the ad-lib feeder introduced. That might explain the drying up of her already paltry milk supply.

David comes over. He has been worried about her too, and agrees that she may not have been eating. Although she hasn't got a temperature, he fears that there may be a low-grade infection, possibly metritis (a womb infection that often follows farrowing), and has treated her with some penicillin. He suggests that he should start to feed her by hand again, in the hope that we can get the milk flowing.

His week has gone fairly well. Litter sizes are up, so it seems increasingly likely that foxes were to blame for the disappointing results over Christmas. But he has found two dead piglets this morning, so perhaps there is still a villain out there. I had noticed that piglets seemed to be taking little notice of the electric fences; this doesn't matter too much when they are young, as they will always return to Mum, but it's not such a good idea if there's a fox about. David admits that the bottom strand is not carrying much current around the newly farrowed sows' pens. He has been short of batteries for a day or two, a situation now rectified after having finally caught a sow who has been reluctant to renew her acquaintance with the boar, and has hogged a whole paddock (and therefore the power unit and battery) to herself all week.

Keeping the miles of electric fencing running is no mean feat. We run each block of pig paddocks off at least one fencer unit and battery, with small solar panels to help keep the batteries charged. Not much use at this time of year. The fences are low – pigs don't jump – and this allows a tractor to drive over them if needs be. The batteries need changing every week in winter, going back to the workshop in the farmyard for recharging. But pigs love to root, as we've discovered, and especially, it seems, around the perimeters of their paddocks. They chuck soil on to the fence line, shorting out the electricity, which means there's a daily task of freeing the wires from the mud before the pigs realize that the fence isn't working and escape. Much as they hate electric shocks, they test it regularly. One of the attributes of a good pig person is their ability to keep the pigs where they are supposed to be! David is usually pretty good on this front, but the winter workload is taking its toll.

Escaped pigs are a nightmare. Fine, as I say, when they are babies; you will often see small gaggles of piglets exploring the tracks and hedgerows (they love hedges), and they will always return home to Mum after their adventuring. But if the gangs of older, weaned pigs get muddled up, they will fight – not such a huge problem outside, as they can get away from each other – and will need separating back into their age groups. A tedious, time-consuming and stressful job. Worst of all is the boars escaping. There are 14 boars, and they work as pairs. The pairs will be brothers, or at least reared with each other from a young age so that they don't fight. If they get together with unfamiliar boars, they will have a real set-to. All that testosterone and competition for the next available 'in season' female. They can kill each other, from heart failure or from injury, if not quickly separated, and separating fighting boars is not a job for the faint-hearted.

I suggest David orders a few more batteries.

As we walk through the pens, we discuss why the weights of the slaughter pigs have been too high this week. We aim for around

80 to 85 kg deadweight, which means their weight when alive should be about 110 kg; a pig loses about 30% of its weight at slaughter, through blood, guts and other internal organs. If they are much heavier than that, the bacon slices are too big for the packs, and there is too much fat on the rashers as well. That's not appreciated by today's customer, and it's not efficient for us, either, as fat takes more calories to produce than lean meat. Then there's an increased risk of 'boar taint' from the male pigs as they get older. This is an unattractive tang that only women can smell, released on cooking, and caused by androgenic hormones.

This is only the second processing week since Christmas, when we had a three-week pause in sending pigs to the abattoir, partly because of the Christmas break, and partly because the weights pre-Christmas had been too low. From one extreme to another. Hopefully we will be back on track next week.

*

Sophie and Dai pop in for coffee and a catch-up later that morning. They have been away for a week skiing, a well-deserved holiday. Those two cover a lot of ground between them, Sophie being in her fourth year at vet school in Nottingham, and now spending quite a bit of her time on placements. Even when it's term time, she dashes back and forth to spend as long as she can on the farm.

Our Sunday-morning get-togethers are a well-established routine. With me away so much, it's often the only chance in the week for a proper catch-up. Henry usually joins us too. Although he and I have been separated for many years, we are firm friends and work together brilliantly as well. He has passed his height and remarkable intelligence to his daughter, though not his love of machinery. Sophie loves animals rather than tractors, as does Dai. Although they currently have a lot of sheep, their particular enthusiasm is for dairy farming. Farming grassland, and turning grass into milk or meat, is what they hope to do more of at Starveall, the proposed site of a new dairy at the southern end of the farm. There are some largely redundant buildings there, and we have applied for planning permission to replace them with a new parlour and with splendid cow housing for the winter months.

My kitchen table is, as ever, covered with newspapers, farming journals and my work papers; I spend many of my weekends labouring away here, and it's always a mess by Sunday! We clear a bit of space, and I put the coffee on. While it bubbles away on the ancient Aga, we reflect on the Oxford conference meat-eating debate, and how the 'eat less' message has suddenly taken hold. The rise in veganism in the last year has been startling, and although some of the propaganda from the most vocal vegan groups has felt deeply unfair and hurtful to the majority of farmers who aim to raise their animals humanely, there's no doubt that many people are starting to rethink their meat habit.

We should hear soon if our planning application for the new dairy at Starveall has been approved. Dai ponders for a moment as to whether this is still the right move for us. It's a huge investment, one which we will be paying off over many years, so if the tide is turning on meat and dairy consumption, we need to be confident we won't be left with a 'stranded asset'. I suggest that dairy is less of an issue than meat; many people will find it hard to turn away from milk in their tea and coffee and their favourite cheese, butter and yogurt, surely, and the environmental problems with dairy farming, done well, are less acute. Compared to meat, dairy farming is a much more carbon-efficient way to produce protein, and dairy-bred calves that are used for beef are a better bet for the environment too than ones reared by suckler cows. The calf is effectively a by-product of the dairy cow; for the same amount of methane (it's the methane which ruminants produce that is a big part of the climate-change problem), a dairy cow produces both lots of milk and a calf, while a suckler cow just produces one beef calf each year, and no milk for us – so that's a lot of methane, land and water (if that's your limiting factor) for just one calf.

Then again, the worst thing about dairy farming is that you need to take the calf from its natural mum when it is just a few days old, so that the milk she produces is captured by the milking machine for us to use, rather than by her youngster. Our Friesian cows will produce enough for both the calf and us, but the difficulties of developing a system which allows this to work have confounded all bar a very few small farmers.

At Eastbrook we have pioneered a way of rearing our calves which seems to me to be the next best to allowing their own mothers to raise them. We have a small herd of 'nurse' cows, older dairy cows who need or deserve an easier life having done their time in the milking herd, or sometimes younger cows who have a health problem, such as mastitis, or lameness, and need a period of recuperation. They each foster two or three calves, as well as keeping their own, so that the calves get the benefits of a 'real' mum, and most cows will have the chance to keep their babies in later life.

Even so, Dai, rugby-playing alpha-male that he is, says he is finding the calf-removal process harder every year. If we are going to go ahead with this new dairy, should we put all our efforts into cracking this problem as part of the project? Even though I know what a tall order this is, I can't help but be pleased that they are thinking along these lines. We have always done things differently here, and this feels to me the kind of challenge that Eastbrook should be up for. With Sophie and Dai's energy, patience and deep expertise with cows, and a clean slate to work from in terms of housing design, we might have a chance of success.

We have a long history of dairy farming at Eastbrook, and we have always done well with it. It's hard work, with all the milking, silage making, bedding, feeding and mucking out that's involved, but it's a regular income, and we pride ourselves on doing a great job, with very low levels of lameness and other problems, and very little antibiotic treatment needed either.

Despite our agreement that the new dairy is still the right way to go for us, it does have implications for the pigs and arable cropping. I've already mentioned that the pigs rotate around much of the land that is close to Starveall, and the crops follow them, so that the ground is out of grass for several years. As the new dairy herd expands, the pigs will be squeezed out, and given what an important part of our business they are at the moment – and that I like them so much – this feels both risky and sad. This is a source of tension between us; I want to keep the pig herd at the size it is now for as long as possible, at least until the dairy is a success and paying off the borrowings it will need to build it, whereas they would prefer to reduce it sooner. I am so aware of the challenges ahead for farms like ours, family farms that may need to fight for survival in the coming years. Just like my piglets!

I remember to take a bag with me when I visit the pigs this afternoon. Although I'm sure it can't be good for the malnourished piglets to suddenly start eating a lot of solid food, I'm also aware that if they don't get some grub soon, they will die. A couple of bags full of pellets should help get them through the night.

They rush up to me this time, seemingly now associating me with food. They are quick learners! They tuck in. What a will to survive this bunch has, against all the odds.

Same again next morning, and I fancy that they may have filled out a fraction over the last 24 hours. It's a bright, cold morning, and many of the more robust piglets in the group are out and about. Molly herself comes up to chew my boots, very bold, and for the first time some of her litter mates are going to the feeder with the sow. David has left one flap on the feeder open, to make it easier for them. Even so, they have to almost get inside the trough, but this isn't going to stop them. They are programmed to eat, these pigs, and will fight their way through any obstacle it seems, to satisfy their hunger.

*

I see that Molly and several of her gang have some blood smeared on their shoulders. Checking them all, I finally find one with a nick in its ear; nothing to worry about, though I can't see what can have caused it. I guess the blood spread on to the others as they jostled to suckle.

I take a walk around the paddocks of older weaned pigs, the ones we call the 'growers'. These are in groups of 80 or so, either all males or all females, with between one and three large metal arcs filled with straw. They start with one arc when they are newly weaned, and then more are added as they grow. There's a veranda of straw outside, which they lie and play on out of the mud, and a straw track to the feed and water too, when it starts to get really wet. They are curious and full of energy, swarming around me, barking and playing with each other. Virtually all pigs of this age are kept indoors on non-organic farms, largely because they are easier to manage, and will convert grain into meat more efficiently if they are not running around and having fun. Our vet, though, is adamant that the reason we have such low levels of respiratory and other health problems is because the pigs are outside. At times we have wondered whether we should try to develop a more 'indoor' approach for winter, to take the pressure off the land and the pig team too. But we keep coming back to the fact that this free-range method keeps them healthy and happy in a way that is very much harder to do inside.

I work all day today until the light is dying, at which point I race up to the field for my last visit to the pigs for a few days.

It's always Tess that I pop in on first, calling out as I approach in order not to startle her. I circle widely to allow her to see me before I stick my head through her door. But it's me who has the shock. With Tess in her arc is a bunch of huge piglets; for a moment I think I'm hallucinating. Then I think that no pigs can grow *that* fast. Then, that they are definitely not her piglets, so whose are they and where on earth are hers? Slightly panicked, I start checking the other arcs. A minute later, I find that a nearby arc, 20 or so metres away, has two sows in it, and about 25 piglets. It's full to bursting, and amongst the throng I spy some of the tiny 11. The sows are lying with their backs to the sides of the arc, udders on display, and there is an unseemly hubbub of piglets large and small fighting for nipples. Like a gambler watching a boxing match, I'm egging my little ones on. They are half or a third of the size of the others, but somehow manage to grasp a teat now and then for a few seconds, to suck a mouthful or two of life-sustaining milk.

How did this happen? I wish that I had been there all day, to observe how these frail piglets had somehow decided to move en masse to another arc. Had instinct or desperation forced one of them to try their luck elsewhere, then maybe told the others that there was another universe out there, with better prospects? Had Tess pushed them out, knowing she was failing them? And why did these two sows decide to get together, to let them in, to give them the chance to feed? We know so little, really, about the lives and abilities and emotions of pigs.

I watch for a while, in a mix of awe and nervousness. So many little lives between the legs of two enormous sows. But interfering could do more harm than good, and I'm not sure what I should do anyway. At least they are getting some food. What will they do now? Will they stay here overnight, or try to return to Tess and the bunch of hoodlums that seem to have taken up residence with her?

It's nearly dark, and I haven't given Dog a walk. The field next door is grassland, and I let Dog out of the truck and walk across towards the badger sett in the hedge on the far side, keeping an eye on the paddock in the gloom. We are halfway across the field when I see a dark shape running away in front of me towards the hedge, towards the large patch of a birdseed mix that we have planted to sustain wildlife over winter. Surely not another fox! We follow in the direction it seemed to be heading; I'm keen to chase it as far away as possible, especially with the prospect of those tiny piglets making the treacherous journey back to their mum in the dark.

I can barely see now. A swarm of lapwings appears from nowhere, screeching and swooping as we walk back towards the truck. I'm walking slowly, eyes pinned on the no man's land between the arcs. Then . . . a flash of white piglet, and another, and another. Three piglets start to trot back towards home. They get to the arc, and stop, bemused, perhaps, by the usurpers inside. They go around the side and back, noses up across the doorway, and then away again. 'Get in, get in!' I'm inwardly screaming. One white blob disappears, then another. Finally, no white blobs. What about the others? Have they made the trek while I was distracted by the fox? Or will they take their chances in the overcrowded arc with the other two sows? It's out of my control. All I know is that this gang are survivors, and desperate to stay that way.

6

A Death in the Family

It's Monday morning and I am in Bristol. But after all the drama of the weekend, I'm anxious to know how the little piglets have fared. Did they all make it home to Tess? Have they continued to go walkabout in search of milk? I try to call David from the Soil Association office, but no response. The phone signal is not always good on the farm, and I will have to wait to find out the news.

It's TB testing week. A year ago we had a major outbreak of this insidious disease, which is driving farmers to distraction in the many parts of the country where it is now endemic. Bovine tuberculosis affects cattle, as the name suggests, and can be transmitted to people via unpasteurized milk. Because of this potential human health risk, albeit a negligible one now that most milk is pasteurized, it is a notifiable disease that successive governments have tried to eradicate, to no avail. The number of cases in cattle each year continues to rise, and it is a deeply divisive problem too, because of the role that wildlife, especially badgers, play in sustaining and transmitting the infection.

Bovine TB can infect many species, and once it has caught hold in an area, it will embed itself in the wildlife, which can then reinfect the cattle. How much reinfection is due to wildlife, and how much is due to undiagnosed cases in cattle which then pass it on to their mates, is a bone of contention. In the opinion of many vets, ever-increasing herd size is a factor, as is the movement of cattle from one herd to another. But there is little doubt that a proportion of cases must be due to transmission from wildlife, with the badger being the probable culprit, as they are the species most likely to come into contact with cows.

This is where it gets emotive. Badgers are a protected species, and very cuddly to look at from a distance. (Up close, they are perfectly ferocious, as I once discovered when trying to help one that had been caught up in some wire.) The idea of culling badgers to prevent the spread of TB has caused public outrage, with campaigners aiming to disrupt the culls in areas where they have been licensed to kill.

As a closed herd – that is, one that does not buy in any cattle except for bulls, which are bought from parts of the country where there is no TB, then tested twice, and quarantined too – it seems almost certain that our outbreak has been caused by badgers. It's hard to be certain, but the 18 cows who suddenly tested positive for the disease in January last year had all been together grazing a pasture that we rarely take cows to, as it's over a mile away from the dairy. There's a long-established sett there, and my hunch is that this was the source of our woes.

Pigs can get TB too, and it's an ever-present risk that our herd will show some signs. It's not a human health risk in pigs, but the main abattoir we use will not accept pigs from a herd that has the disease. There are so few abattoirs that have an export licence, and who are prepared to handle the relatively small numbers of pigs we send each week, our business would have another mountain to climb if we had to find another suitable abattoir.

Once a farm has an outbreak, all cattle over 12 weeks of age must be tested every 60 days. That's one hell of a job. We have 200 milking cows, some in the herd, some on holiday between lactations on the turnips near the pigs. Then there's the nurse cows, some 45 of them, with 120 or so babies of various ages, in two locations. The dairy heifers with the bull, another 49. Young beef cattle at Starveall, 68 in total. There are also 27 older beef cattle in the Eastbrook yard, three Angus bulls in with the cows, and the two new Herefords (Branston and Pickle) still in quarantine. All need to go through a handling system so that they can be scratched with the inoculum by the vet. Then, three days later, they need to be caught again, to see whether they have reacted to the scratch, in which case they will be deemed to have failed, and will be culled.

So our week looks like this: all day Monday and Tuesday, the cattle are caught and scratched. This takes a team of four staff, plus the vet, so there's little time for any other work. Wednesday is a day 'off', so a frantically busy one, doing all the jobs that have to be done, that there hasn't been time for. Then all day Thursday and Friday, the moment of truth as the tests are read.

Today is Wednesday, the one day of the week when the team

are not spending the whole time catching cattle. I am at home, my last day before a fortnight's break, our annual retreat from the real world. It is a chance to finish the things in all the compartments of my life that need to be completed before we leave. And to spend a last few hours with my pigs, of course.

*

I spend the morning reviewing progress on some of the bits of the business that I tend to neglect somewhat. Or rather, businesses that operate from here, but aren't under our direct control – which makes life doubly difficult. We have always liked the idea that a large farm like this should sustain as many people and activities as possible, and yet there's only so much that it's possible to manage ourselves without going completely mad, especially when I am away for much of the time. So when individuals turn up with a good idea that might fit with what we currently do, we have tried to help. But it hasn't always been successful, I have to admit.

One such person is Claudio, the larger than life, good-looking son of Italian cheese makers. He wanted to start a business making organic mozzarella and burrata from cows' milk rather than from the traditional buffalo milk. Given that we want to start adding value to our own milk, so that we are less at the mercy of swings in the organic milk price – especially likely if lots more farmers move to organic dairying so that supply outstrips demand – this seemed like a low-risk way of diversifying our market.

So we looked around the farm with Claudio for somewhere suitable to convert into a processing dairy. His eyes lit up when we showed him the little thatched building just behind the farmhouse, which we used years ago as a butchery, and indeed, was then used for cheese for a while by some folk in the village. It is small, but there's a cold store already attached, and while it needed a bit of refurbishment, it was a relatively cheap and cheerful way for him to get started. The problem is that the business is growing quickly, and there is not enough room in the Cheese Shack, as he calls it, for him to expand; added to which the drains cannot cope with the volumes of water he uses to wash down the plant each day, so we have to remove this in tanks. This is an unanticipated chore for the farm that we could do without, and while I'm delighted by the success of this enterprise, and its rather cocky leader, we need to find him a more suitable site as soon as we can. Meanwhile, his delicious cheeses are a hit in the pub and Chop House, and it's great to be able to feed people an artisan product which hasn't left the village from start to finish.

We have started another venture with a brilliant wildlife photographer, Elliott, who, having got to know the farm when he was commissioned by us to do some pictures for our website, was bowled over by the range and number of birds and mammals here, and asked if we would consider installing some hides so that he could rent them out, and teach from them too. We liked this idea a lot, especially as it should provide another great reason for people to want to stay with us, and so now we have six hides around the farm, positioned next to badger hides, ponds, and the Starveall cottage where little owls regularly nest. The feedback from Elliott's clients has been overwhelmingly positive, but I've not seen him around for a few weeks; I do hope he hasn't been eaten by pigs!

*

I force myself to get through all these duties and paperwork, but I'm on tenterhooks to discover how the survival tactics of Tess's litter are working out. Finally feeling that I've earned a break, I drive up to the field and approach their paddock, so familiar to me now that I know every arc (there's an assortment of them, some wooden, some metal, of various ages) as well as their inhabitants. But something's changed. Some of the arcs seem to be missing, and in the centre of the paddock are three very large huts with a stack of bales to one side. Inside these are five sows, in deep fresh straw, with a vast array of piglets.

David comes over. Having seen the way Tess's piglets were roaming about, trying to find surrogate mums with more milk, he has decided to encourage them to multi-suckle by replacing all the individual arcs with three communal ones. Brilliant thinking. I'm delighted; this should give the weak ones a much better chance of securing the milk they need.

He's built a protective straw-bale wall to cut the wind, and bedded up outside the huts too. It looks very cosy, and I sit down in the shelter, with piglets all around. He hasn't removed Tess's arc yet, and she is still there with some of her youngsters. Others from her litter have already made the move across to their new haven, and are trying to feed from their 'aunts', wrestling their stronger cousins as best they can. Despite the jostling within the deep straw of the arcs, it's a peaceful moment, tucked up out of the wind. I look around the field from my protected hideaway, and am struck again by the number and variety of birds that are using the pig field as their winter refuge. The pied wagtails, carefully following the pigs as they root the ground. The redwings and fieldfares that swing along the roadside, moving between the bushes, and then settling on the undersown stubbles on the land around. The gulls and crows of course – we sometimes wish that there were fewer of them – and the extraordinary numbers of migrating starlings that provide an almost constant aerial display of stunning beauty, creating a kaleidoscope of fluid formations as they rise and fall. The lapwings are not in such great numbers, maybe only 400 or 500 compared to the tens of thousands of starlings, but they too provide an exquisite understorey of shapes. Right now, both are in flight, the starlings high and the lapwings beneath them. Despite my farmerly concern about how much of the pigs feed they are stealing to sustain their lives and artistry, I begrudge them not a bit of it at this moment; their beauty is ample reward.

Beauty is too frequently neglected and unrewarded by our often utilitarian view of the land, even though it's what many people value most. The awe that the land and its creatures inspires lifts our sights beyond day-to-day drudgery, much as music can, leaving us with a sense of our place in the world, as part of nature, part of the flow of life that relentlessly carries our genes, constantly evolving, constantly adapting, like a river towards the sea. As drops in that river, we are individually insignificant, but humanity's great gift is consciousness, of being able to watch for a brief moment this extraordinary flow of life, a miracle in a seemingly dead universe. Momentarily we are snowflakes, each one unique as we drift towards the earth, with a perspective that few, if any, other species are granted.

This view is a partial one, however, so that this fragment of time when we are individually aware is also a danger. We tend to believe that the world, or at least the countryside, should be as we first knew it, without a sense of what came before. So I have grown up believing that these chalk uplands are as they were meant to be, open landscapes, grazing sheep and growing barley – but we have made them this way, albeit many hundreds, possibly thousands, of years ago now. Left to their own devices, scrub and woodland would regenerate, depending on the balance of other wild species that might make their home here. Most of our land in the UK has been shaped by humanity, and by our domesticated farm animals, which have replaced the wild ones.

Beauty, then, is not just in the eye of the beholder, but is also only skin deep. What can seem a glorious sweep of green fields, maybe aesthetically enhanced by a carefully positioned copse or hedgerow, can disguise a dead soil, reduced to a largely inert material by decades of arable cultivation, and no return of organic matter to feed its starving earthworms. That deep greenness too, can be deceptive, the result of over-use of manufactured nitrogen, forcing the crop to grow faster than it should, weakening it so that it succumbs to disease, or pest attack. The chemicals we apply to cure these ails may then mean that even that copse or hedgerow, meant as a refuge for beleaguered wildlife, can become a death trap, especially for insects, as those chemicals leach into the more natural vegetation. All is not as it seems. We need to look below the surface, to ensure that our instinctive empathy with the type of landscape that we recognize and perhaps feel safe in, is not just a hard-wired response which stems from our lack of a deeper understanding and a thoughtful imagination.

People sometimes ask me how they can tell if a farm is run along organic lines. In this pig field, it would be obvious to them if they knew anything at all about pigs and the way they are normally kept. But it can be more difficult if you are gazing out across a field of grass, for instance; it can all look rather the same at first sight. But take a few paces into the field, and the differences become immediately apparent. Look down. An organic pasture should have plenty of clovers, fixing the nitrogen that will power the growth of the grasses, and there will be a much wider range of other species too – maybe, as in many of our leys, bird's foot trefoil, some chicory on our heavier clay fields, or salad burnet and plantains. Up here on the chalk, sainfoin thrives alongside the white clovers. There are several species of grass, cocksfoot and timothy, which can withstand drought, as well as the ryegrasses, which provide high-quality fodder for the cattle and sheep. And in very long-term pastures, all sorts of native species will start to appear if the management allows, including, of course, some that farmers hate, like docks, nettles and thistles!

So it takes a little knowledge of what to look for before that beauty can be taken at face value. If I had to advise people to look for one thing, though, in almost any circumstance – not just on the land – then I would say 'diversity', the root of resilience and, happily, often of beauty too.

Shaking myself from this reverie, I leave David to finish the reorganization, and continue the endless list of tasks that need attention. Going on holiday creates an unmoveable deadline, absorbing me until late in the day. Tim comes in and surveys the mounds of domestic chaos around the place, and me still focused on my laptop. 'Have you not packed yet?' he exclaims, though of course he hasn't even thought about it yet either and, knowing him, he won't until an hour before we leave. He starts moving the chaos around in a not very determined manner, clearly in the hope that I will do something about it all, but it's nearly dark, and I must go to say goodbye to the pigs.

All the individual arcs are gone now, just patches of bedding across the paddock showing where they had been, and every litter of piglets is ensconced in their rather heavenly new home, rummaging around in the straw, playing and rooting. The size difference between the malnourished 11 and the rest is even more noticeable now that they are all together, but they seem healthy enough and perfectly able to hold their own against their more advanced cousins. They seem to be feeding from a variety of sows, running between the three large arcs, from one group of mums to another. I could stay forever, watching them, loving to see the way the gang has amalgamated so quickly, loving them. It's so hard to leave them for a fortnight, but at least I'm feeling far more confident about their chances now.

*

Inexplicably, I wake at 3 a.m., the next morning. We must leave the house before dawn and so rather than lie awake, I sneak out of bed and downstairs for coffee. Moonlight streams into the kitchen. On a whim I pull on my super-warm farm coat, waterproof trousers and boots over my pyjamas, and head to the pig field. The air is crisp and cold; breathing is like drinking ice-chilled water. I park in the gateway and, torch in hand, trudge through the lightly frozen mud towards my gang.

Even without a flashlight, it's easy to see once my eyes adjust. The pigs will think I'm mad, but I just have an urge to see them once more, settled into their new commune. I don't even call out as I approach; it seems almost sacrilegious to break the absolute silence of the still night.

As I approach, an eye or two flickers, and an ear or two twitches. They know that I'm there, but seem sanguine about their night-time visitor. The majority of the piglets are in the middle arc, the largest one, with three sows. To the left, Mrs Messy Bed, the prick-eared sow who couldn't even make a decent nest, and one other are gently snoring, undisturbed by any youngsters. They seem to have decided that the new arrangements have let them off maternal duties! To the right, three more sows, backs to the walls, and a cluster of the smaller piglets, including many of the underfed 11 who are sleeping soundly, it seems contentedly full, at last. There's room for me too, and I creep inside. There's an alarmed grunt or two, some snuffling and re-positioning, and then we all settle down in the warmth of the straw and body heat.

It's deeply peaceful. I watch them for a while, the micro-movements, the odd grunting adjustments, the mass of breathing life. I doze myself for a few minutes and awaken with a cramped leg. I shuffle around in an attempt to get more comfortable, and two of the sows raise their heads in surprise; they perhaps hadn't noticed me before. The third is motionless, an arm's length away to my right. It is Tess, lying peacefully surrounded by piglets. So still. Too still. I reach out to touch her, and find her cooling, stiffening. She is dead.

I sit for a while longer, both sad and bemused. She looks so at ease. How extraordinary that she had struggled on, desperately trying to care for her young, and now that they are safe has given up the ghost. There is a strange serenity in the moment.

7

My, How You Have Grown!

6–12 February

We have escaped the winter to recharge our batteries in paradise. Ko Phra Thong is a remote, unspoiled island in the Andaman Sea, recently designated a nature reserve. It's easy to understand why. Fringed with forest, its interior is savannah, home to over 125 species of birds, while the miles of empty beach make it a refuge for turtles. It was hugely damaged by the tsunami in 2005, but has recovered quickly, and even the coral is coming back. We live simply but comfortably here, sleeping to the sound of waves in our stilt house, open to the elements bar a mosquito net. Having discovered this haven quite by accident 12 years ago, we return every year, despite our guilt about unnecessary flying.

Valerie and Jocken are our hosts, and they have made every effort to maintain and enhance this paradise. They employ their Thai staff all year, even though visitors only come between

November and April, and have started to grow what food they can here; tricky on a giant sand dune. For a while they even had pigs. Just three, in a large compound from which they regularly escaped, so that we would often meet a sow or two trundling down the sand tracks. We got to know them well.

On our last night two years ago, Jocken and Valerie invited us for a special farewell dinner. Jocken had been in the kitchen all day, preparing dishes from one of the three pigs that we had become so attached to. Delicious, but I found myself mourning the pig. Attachment is a dangerous thing.

There are no pigs here now, and Valerie has become a vegetarian, somewhat scarred by the experience of slaughtering animals that they had become close to. But their legacy lives on, in the relatively fertile soil that they created in their compound, now growing papaya, lemons and some vegetables. Animals can transform arid land, building organic matter, nutrients, and biological life into soils far faster than is possible without them. I have seen the same on my friend's farm in Jamaica. The thin gravel soils are best reinvigorated with animal manures, and since she has started to keep her pigs free range, rather than inside, and now with chickens being folded across the land too, the yields of her citrus fruits and coffee have risen sharply. In all of the debates about meat-eating, the role animals can play in building soil fertility in fragile lands – as well as providing much-needed protein for peoples less privileged than us – is rarely mentioned.

One of the books I've been reading here is the story of Knepp Castle, the large Sussex estate that has been rewilded by the pioneering Burrells. I went to visit Knepp last year. In a few years, this heavy-clay land has transformed from arable farm to wildlife haven, with a succession of trees and shrubs already established. They have introduced animals – native ponies, cattle and Tamworth pigs – into the burgeoning woodland – at low levels to start with, otherwise they will prevent the regrowth.

Having released the animals into such an expanse, their approach is largely hands off. They check on them regularly, if they can find them, but allow them to forage as they will, to take their chances in the natural environment. We met the Tamworths on my visit. A small family, a sow with a teenage pig from an earlier litter, plus a couple of fat little piglets from her most recent batch. She will have had more than this, but as with all wild pigs, many will be lost to predators. Animals that have large numbers of young lose most of them. The turtles on this Thai island hatch hundreds of babies, only a few of which will make it to adulthood, the rest providing sustenance for the wider food chain. Wherever you are in the world, the same rules of nature apply: everybody eats everybody else, and of all the young that are born in the wild, only a few will make it to adulthood.

As farmers, we protect the vulnerable young, so that we, not a fox, may finally eat them. What would the pig make of this deal?

*

We arrive home early on a cold, clear morning, and I can't wait to see how my pigs are doing after being away from them for more than two weeks. What a welcome! The sows and youngsters swarm around me, almost knocking me over as they come up for a scratch. I'm taken aback by their enthusiasm. Do they remember me?

Molly and her siblings must weigh over 15 kg now, solid balls of muscle. It's an absolutely freezing morning, but they seem entirely impervious, romping around with abandon. Most of Tess's piglets have grown, though they are still half the size of the other litters, but three of them are real runts. They have little body fat and the swollen stomachs so typical of the underfed. I am very worried about them.

This is the hardest time of year on the farm, and especially in the pig field. The days are still short, the weather abysmal, the winter routines seem never-ending, and it feels as though spring is a long way off. It takes special people to work outside all day in these conditions, however much they care for the pigs, and by now, the cold and mud have taken their toll. David's back is playing up. Clive, who helps with the pigs if we are short of manpower, has been on holiday and then very unwell. Looking around the field I can see that we are getting behind on some routine tasks. The new farrowing paddocks have not been divided up, there are arcs that need repairing, and I suspect that tasks like pregnancy testing have gone by the wayside for the moment.

When I meet up with David, he confirms that things have got a bit behind due to the weather problems and staff shortages He has missed a week of weaning, which in turn means that he's missed a week of serving – that's farmer's speak for mating – sows too. It's not a welfare problem; the piglets will be perfectly happy to have another week with Mum, though she may be looking forward to some peace and quiet by now, but it gets everything out of sync. The whole system works to a beat. Every week we should have eight farrowings, eight sows being weaned, eight sows becoming pregnant, roughly 80 pigs going for slaughter. We have the right number of arcs, feeders and troughs for this routine; if it gets out of rhythm, then we can suddenly find ourselves with problems, such as not enough farrowing arcs for the bulge of births that will result from an absence of them in previous weeks.

Then there are other jobs that can slip at difficult times, but which will come back to haunt us in the end. Every piglet is vaccinated at four weeks, and again at weaning, against a rather nasty 'wasting' disease that can turn healthy pigs into skeletons, if indeed they survive. Since we started vaccinating, we have hardly seen a case, and the same with meningitis. For many years we would lose a few pigs each spring and autumn to this distressing condition. Then our brilliant vets told us that a company had started to make short runs of vaccine, using the bacteria from farms, so that the inoculation was specific to the bug in our own pigs. This has worked exceptionally well, and our post-weaning mortality has dropped from a very respectable 1.2% to an incredibly good 0.5%. I'm keen, therefore, to make sure that we don't let the programme fall behind. Molly and her gang should have had their first jab a couple of weeks ago, but the weather and workload has meant that this hasn't happened yet.

One of the jobs that adds to the never-ending list of things to do in winter is water bowsering. All the paddocks are supplied with water from a network of blue plastic pipes. Because the pigs move over so much land, these pipes run overground, so freeze up fairly readily. When this happens, we need to take water to each paddock in a tank, piping it into the troughs a couple of times a day. Pigs need plenty of water, as they are eating dry feed, and will soon run into trouble if the supply fails. It can take one of the guys half of the day just to do this, leaving less time for all the other jobs. When it's raining rather than freezing, it's bedding-up that can take the time; straw gets sodden, and regular new supplies are vital to keep the pigs warm and dry.

I'm frozen myself, and exhausted from 24 hours of non-stop travel, so I retreat to unpack, and start on the mountain of emails and post that has accumulated in my absence. Amongst the heap of letters is the written confirmation of our bovine TB test. I had heard the news while we were away, but it's good to see it in writing: just one 'inconclusive' cow. This means that the vets cannot be sure whether she has reacted to the inoculum or not, so she will need to be re-tested in a couple of months' time. That's a pretty good result considering that last year we lost so many cows. I plough on for a few hours, and pop to the farm office to say hello to the team there. But I can't resist going back to spend more time with the gang later, even though they seem much less excited to see me, and largely carry on with their lives as if I'm not there.

The disparity in size between the majority of the group and Tess's orphans is clearly a problem. They have had such a tough start in life; a big litter from a sow who provided little milk, scavenging where they have been able to from other sows, and then losing their mother too. The little ones scurry back and forth, burning up ever more energy trying to grab a bit of feed, either from a sow or from the feeder. They know where the feed is, but struggle to lift the metal lids. When their robust pen mates flip up the lids with ease, the runts sneak in alongside, but are knocked out of the way before they can snatch more than a few crumbs. There's not much 'after you' going on here.

I prop up one flap on the far side of the feeder in the hope that it will allow the small ones to get a look-in. One of them finds it almost straight away, and dives in. He is small enough to get completely into the feeder, with only his bum protruding. Within five minutes, though, the others find his easy access, and bunt him out of the way.

Over the next few days I grab what time I can with the group, trying to work out what is best to do with the laggards. There is one that I am especially worried about. Pot-bellied, with no fat at all over his hunched spine, he looks as though he could easily die. It's hard to know how best to help piglets like this. At weaning, they will go into 'special measures', a small group of disadvantaged pigs that can be nurtured until they catch up, but it's going to be at least two weeks until this gang are weaned. Are they better off with the family they know, stealing whatever food they can, or would they thrive more away from the sows and the competition of their robust mates?

The situation is not helped by the simply atrocious weather conditions on Thursday and Friday. It is bitterly cold, with wind and rain too. Tony has inconveniently taken two days' holiday, and David struggles alone in conditions that would defeat any normal person. In these circumstances, we would normally find someone else on the farm to help, but Clive has been ill for three weeks now, leaving the whole team under pressure. David had planned to be catching up with vaccinations, but when I see him late on Friday he explains that all of this has had to go by the wayside. This in itself has led to a degree of chaos. He had removed the feed hoppers in the paddock of pigs to be vaccinated, as the only way to catch them all is to lure them into a hurdled pen with feed. Having given up hope of completing the task today, he has just reinstated the feed hopper, so all the sows and piglets are hungrily competing for feed spaces. The bangs of feed lids, and squawks from the piglets as selfish sows push them out of the way, add to the clamour of the starlings flying low over the field in the force eight gale. Again, I am reminded of how competitive pigs are. They have little altruism, it seems to me, especially where food is concerned. They are born to eat, and will let little stand in their way, not even their own piglets.

They are the only pigs feeding in the twilight. David, still surprisingly cheerful despite his appalling day, remarks how the whole field of pigs responds to the weather. No one ventures out while the wind is howling and the rain is lashing down, but within moments of any respite in the conditions, the pigs sprint to feed and drink, then dash back again as soon as the wind and rain picks back up. I tell him to get home and warm, and to be honest, almost chicken out myself from my evening tour. It's hard to stand up, my boots are balls of mud, and I'm grateful for waterproof trousers as well as jacket, thermals and hoodie. Not the moment for a *Vogue* shoot!

As David predicted, Molly and her gang are all tucked up inside. I duck into the shelter of their arcs, somewhat startling them; it's such a relief to be out of the wind. The arc is toasty warm, at least by comparison, and the piglets are mostly buried in deep yellow straw, their own giant duvet. The sows ignore me after their initial surprise, but the Saddleback boar and the big-bottomed brown Duroc throwback extricate themselves and come to investigate. I sit in the doorway, they chew my boots and submit to a bit of back scratching. It's hard to leave, especially when it means facing the storm again. When I finally do head back, I'm pleased to see that the sows in the not-quite-vaccinated paddock have taken their fill of feed, and left the piglets a chance to sate their appetites before the night sets in.

*

By Saturday the weather is calmer, and it's easier to see the whole group properly. The little runts are still very poorly, and I feel we have to take action. Given how attached I have become to this gang, and how busy David and Tony are trying to catch up after the last few days of storm, I would rather like to have a go at rearing them ourselves. But I'm away much of the time, so this would need to be Tim's project, and one that is well outside his core skill set.

My life and his are usually in somewhat parallel universes. I am up ultra-early most of the time, to beat the traffic to wherever I'm going, usually Bristol or London, or if I'm at home, to walk the farm at first light before the day's work begins. Tim often works late, in the Oak or the Chop House, getting home long after I've crashed out. Some days, the main moments we get together are over early-morning coffee and juice; it's not the best time for a tricky conversation, but it will have to do.

This particular Sunday morning, over that early coffee in the gloom of the not-yet dawn, my tentative suggestion that we might add to our chaotic lives by taking in three fairly pathetic runts goes down better than I expected. I can see that he is trying to hide a grin. It turns out that he had long hankered after a pig or two, having been taken with one adopted by a roguish farming friend of his, who even took it to the pub with him where it performed party tricks for crisps and diluted beer. Although having run our meat-marketing business for over 20 years, and been responsible for the humane slaughter of many thousands of pigs, it was going to be a bit of a volte-face to spend hours of his time keeping them alive. 'But I don't know anything about pig keeping . . . and what's going to happen to them if they do survive? And how will I feel if they don't?' he blusters, while already looking for a spade to start clearing out the old dog kennel, which seems like the best initial home.

So now we have three small, malnourished, rather feeble-looking pigs living in our brick dog kennel and yard, with Tim on a steep learning curve. But they have doubled in size since they arrived here so he must be doing something right!

8

Paternity Leave: An Excerpt from Tim's Diary

Helen's early-Sunday-morning observation of her pigs that week led to the three weakest piglets being singled out for rescue. I'd taken some of our hotel guests up there the same morning, and the pigs were obviously in trouble, engaged in an unequal fight for food and warmth. Survival of the fittest among some completely unsentimental siblings who were already stronger, bigger and noisier, meant that these ones were not going to survive without intervention. That was the simple logic.

I collected them from David the pig manager at 3.15 p.m. the same day by the gate to the pig field. 'You're both mad' was clearly written all over his face. The piglets were truly pathetic; for their age, nearly seven weeks, they were about one third the size they should have been. Two of them had the slightly coarse hair often seen on a malnourished pig; the same two had the slightly rounded body shape and hunched back that often signifies the runt of the litter. There was no cushion of fat or flesh to mask their vertebrae, which stuck out along their backs. And one of them had very crusty eyes, making them look almost permanently closed. The third one was just, well, very small, but in slightly better body condition – finer hair, no hunched appearance. They probably weighed about 3–4 kg each; their siblings at this stage, and the other young pigs they were competing with in that same paddock, would have been 10–12 kg, possibly more. In the back of the Land Rover, they hid under the back seats and peed everywhere, silently. Until I tried to get them out, when their squeals must have alarmed the neighbours.

Carrying them like you carry a baby, I plopped them one by one gently on to the straw-strewn kennel. The outside temperature that day was just below freezing. They moved fast through the straw, as far away from me as possible, and buried themselves, with their heads towards the wall. Two side by side, and one on top in the middle, an interesting triangle of pink bottoms viewed from behind. Serious early bonding between themselves for safety and warmth. But not a single curly tail on view – instead they were lank and hanging, the sign of a physically unhappy pig. That first curly tail was a milestone to be looked forward to, I thought. How long would it take?

Three or four times a week Claudio and his Italian team are making artisan mozzarella and burrata with Helen's organic milk in the small thatched building at the back of the farmhouse. One of the benefits for the house – the only one I've come across so far, apart from the occasional fresh cheese – is that making mozzarella produces huge amounts of whey, the residue from the milk, when they've used all the fat and some of the proteins from it. Wiltshire was once a huge centre for dairy production and pig farming – they go together because pigs love whey. A cheap source of good food, once upon a time, though the gym bunnies seems to be addicted to it now. So there's whey 25 yards from the pig kennel – they can have that, as much as they want, and David has provided a first bag of organic pig nuts for them. I dig out some old dog-food bowls, and these three cowering, shocked piglets have the basis of their diet for the next few weeks, along with some water. Of course, they ignore this feed offer over the first few hours, too shocked to want to move around their new accommodation.

It was dark by now, and everything pointed to the temperature dropping overnight to minus five, or close. It was one of those brilliant winter nights, an almost-black blue sky, everything around us completely still, and our piglets with no other pigs to keep them warm. Even without their mother, the community of pigs that Helen and David had created in their three large arcs a mile from here meant a non-stop supply of body warmth. Here, there were three small, ailing piglets, and no big fat sows acting as central-heating radiators. I hinted to Helen that we ought to bring them inside into our scruffy utility room, but she reckoned that another change, another move, another different location, might be enough of a shock to finish one or two of them off. She suggested a hot-water bottle, and I happily agreed. We found a tough old plastic detergent bottle, put some medium-hot water in it and wrapped it tightly in an old pink towel. I ventured outside by the light of my old iPhone, and tried to put it somewhere usefully near them. They were buried, but they grunted in alarm as I leaned over the impromptu gate I'd erected. I dropped it close to them and hoped for the best as I whispered 'Goodnight, babies' to my new, weak and scared charges, before heading back inside. 'Why do we worry about them so much now they're 25 feet away from us?' I mused. 'I wouldn't be thinking about them if they were a mile away in their paddock, even though they would still be struggling.'

Any concerns I had about not sleeping because these piglets might be freezing to death close by, came to nought, and the first thing I recall, despite being determined to check them at first light, was Helen, remarkably, bringing me a coffee in bed and saying, 'No deaths overnight.' I leaped out of bed and struggled to get something warm on to allow me outside without freezing to death. I wanted to see for myself. First thing I saw in the half-light was a grubby pink mound, completely still and lifeless. I reeled. That is dead, I thought. I leaned in and touched it – it was the pink towel wrapped round the improvised hot-water bottle, of course. Idiot me, I thought, why couldn't I have used a blue towel? Elsewhere, some feed had been eaten, a water bowl overturned, whey drunk, and what was left was full of dirty straw. They'd been standing in it. Honestly! No sign of the pigs except for a large straw-covered hump at the back of the kennel – a hump that was moving rhythmically as they breathed. As I spoke to them their heart rates increased, and the hump moved up and down more jerkily. When I shut up, it visibly calmed down.

Later that day I went to spend some time with them, though I'm not sure they were keen on that, since they moved to the opposite end of the kennel, and buried themselves again. I would extend the hand of friendship, and they would run for their lives. When I managed to touch one of them, it squealed loudly. I backed off. I crept up on them an hour later to find them feeding hungrily. And, amazingly, one of the tails was showing a slight curl. Real visible progress, but the tail straightened again when it saw me watching. The most runtish of the piglets was the most greedy with the whey, while the other two concentrated on the water and dry pig feed.

Straw itself is a perfect medium for them to start to display their rooting instinct, and their active waking time, only a small part of the day at this stage, when not feeding or peeing or pooing, was spent simply digging their noses into the loose straw, piling it up in front of them like a snowplough, not that there was anything really interesting or edible in it for them. It's just what they do. Later in the week, Helen plonked a pile of woodchips, earth and turf into their new outside run, and within minutes they were digging in it. A much more likely source of worms and bugs, and importantly, a peck of dirt to get their antibodies active, starting to provide the resilience they need.

Day two saw the arrival of some slightly grubby bottoms, to be honest. The shock of change must have been too much for them. And then I thought a treat for them would be some old homemade bread mixed with warm water and a drop of milk. A bit like a porridge. I told Helen and she said it was too soon for much wheat. Keep them on their dry pig feed and whey. Get their stomachs used to this change of lifestyle. Like us, their guts can often be the first obvious physical response to dramatic change, so keep it simple.

But over the next few days, their confidence grew and their bottoms started to be prettier to look at. They still buried themselves in straw when sleeping, and preferred to stay there when I popped my head through the gate, or stepped inside to feed them, but they were not always at the back of the kennel. They seemed happier to sleep facing the door, and the one with the crusty eyes was now in a much better state, eyes clear and wide open. The spines on the runty twosome were less pronounced as the flesh and fat around them began to respond to the generous nutrients we were providing in an uncompetitive environment. There was the occasional flash of a curly tail, only fleeting – but very rewarding to see.

They now explored all the space we were providing. We took away the impromptu gate, allowing them access to a small open run where we put the woodchips, earth and straw. Within a minute of this barrier disappearing, they were out, rooting, exploring, and finding quickly where their new feed bowls were. Late one evening, way past dark, I went to top up their water and had to go back into the house for something. I thought I'd closed the main gate, so was surprised when I came back to find two of our charges completely out of their run and pen, and wandering slightly nervously round our back yard. It took some soothing words and gentle coaxing to get them back in. They knew where they had to go, they could hear the other member of the trio grunting as if to say, 'Where are you, what's going on out there?' But I had to show them the way. I could just about steer them with my hands, though they still preferred not to be touched. Sam, my son, visiting from his home in Bristol, suggested throwing some of my old clothes into the pen so they could get used to my smell. I was loath to part with my holey old jumpers. I'd try to get to know them, and them me, by being there, I decided.

Something all pigs seem to love is a good scratch. From the smallest to the largest, it seems there is little they enjoy more than to feel something abrasive against their skin. One of the reasons that proper pig housing, like the wood and tin arcs we have on the farm, lasts such a short time is that they can't put up with a 350 kg sow giving herself a good scratch against the edges ten times a day.

Little pigs are no different, so our three, from halfway through the first week, were rubbing themselves against the brickwork, and occasionally using their back legs to give their ears a good rub too, just like a dog. Giving a pig a scratch seemed a good way of ingratiating myself with them, being useful in their eyes, rather than just wanting to bond, but not being sure how. One of them was always the least wary of me, so as it walked confidently past me on day four when I went in with my bucket of whey, I leaned over and rubbed my fingernails down its spine. Instead of squealing and running for cover, it stopped. So I did it again. Put my bucket down and did it again. I crouched beside it. It looked at me. I scratched its belly. It rolled part way on to its side and exposed its whole belly to me. I rubbed and scratched vigorously, avoiding any sudden movements. Its legs gave way and it flopped completely on the floor of the pen. It wanted more. All over its tummy, behind its front legs, and on top of its head. It grunted – it really grunted – in what I assumed was pleasure. When I pulled away, partly because my own legs were aching from crouching, it stayed exactly where it was, and effectively asked for more. This was great progress. I could have spent an hour doing it and it would not have moved, just grunted happily. And of course, this mutually indulgent bonding was useful for something else – I had discovered that my biggest runt was a boy!

At the end of week one, they are now prepared to look me fully in the eye, and me them. They have caused me more worry, day and night, than anything else I have undertaken since the birth of my own children. I have gone to bed worrying about three pigs, and have woken up wondering if they're still alive. Most odd, I have to say.

9

In Praise of Trees

14 February

I'm still debating whether we have done the right thing in picking up the three most struggling piglets to rear at home. The other eight are still with the group, and it will be interesting to see how they compare in a few weeks' time. Though, of course, we have lifted the very weakest; most of the others are making good progress, even if still smaller than their cousins.

It's mid-February now, and the days are already much longer. On Valentine's Day it really does seem as though the world is waking up. There is the beginning of a dawn chorus, and the five hares I put up on my walk with Dog at Lower Farm this morning seem to be thinking about boxing. They take off in groups and are fairly sanguine about our presence, but don't go as far as the high jinks we will see in a month or so. Around the pig field, skylarks take to the air, specks of music against bright blue.

We are due to get the verdict on our dairy planning application today, and Dai is on tenterhooks, calling into the office at every opportunity to see if there is news. Late in the day, the email arrives. Success! Now we have a whole heap of work to do getting quotes, triple-checking our financial model, talking to our landlords, and seeing whether we will be allowed to bring in young heifers to populate the unit next spring. The last on this list depends on whether the cow who tested as 'inconclusive' at the last TB test goes clear at her March re-test. If she does, Defra have confirmed that we will be considered free of TB, and so able to buy in animals. If not, we will need to apply for a licence to move stock on to the farm, and we have no idea whether this will be granted.

We meet to agree a way of dividing up the tasks. Dai will start looking for heifers, and he and Henry will get quotes for the building works. Henry is already starting to demolish the dilapidated shed where the parlour will go; it's unsafe, so needs to come down, whether or not we go ahead.

The pig unit has started to move into Flaxfield, adjacent to the now rather muddy Broad Gap where they have spent the winter. There are three paddocks of dry sows with boars, and a block of sows due to farrow too. It makes life so much easier when all the pigs are in one area. We used to keep the breeding herd separate from the growing pigs, but David and Tony can coordinate their work better when they are together. It also means that we get across fields more quickly, and can then clear them for the next crop in a more efficient fashion.

The field is a hive of activity with the weather so mild. It's perfect conditions for rooting, and boy, our gang have gotten seriously carried away! Three of the sows seem intent on digging to Australia; there is a series of pits almost three feet deep in the north-west corner of their paddock. They are like mini-bulldozers, pushing up mounds of soil, chomping noisily on whatever it is they are finding to eat.

The piglets prefer straw to nose about in. There is an expanse of the stuff in front of their arcs, and they seem to eat quite a bit of it, as well as digging through it. When I arrive, they switch their attentions to my boots. What is the eternal fascination that pigs have for boots? They throng round, nosing then chewing, pulling at the laces. I can take so much of it, but as they get bolder and bitier, there's a risk I might have to walk home barefoot.

There are a few of the piglets that stand out from the crowd. Molly, of course, with her black head and splodge on her bottom, one of the biggest of the gang, and, if I'm honest, a bit of a thug with the other piglets, though reticent with me. The two pure-bred Saddlebacks, one boar, one gilt, are both well grown and looking rather distinguished against the assortment of white and mottled pigs in this group. There are also two poor piglets who have typical Saddleback markings, but for some reason are not thriving as well as the rest. I think they are Messy Bed's, the sow who seems to lack any notion of competent parenting. And of course, Tess's piglets are always apparent, still lagging behind the others. My favourite, though, if I'm allowed one, is the brown pig that is best showing her Duroc genes. She is chunky, both across her shoulders and her bum; she has buttocks that would light up the eyes of a ham maker. She is often first out to greet me and, like her mum, has a playfulness about her. Having come up to do the boot-chewing thing, she will prance away, kicking up her heels, then stop dead. And then come back for more.

While I am happily messing about with my piglets, and admiring the trench-building efforts, I see David trying to move a sow in the adjacent paddock. The lack of dividing wires is coming back to haunt him; two sows have camped together and he is trying to split them up. I go to help. We get one of the sows out, and he closes the door to the arc to stop her returning. Then we encourage her across the paddock to find an empty arc for the night.

To move pigs you need boards. Unlike cattle and sheep, they will not easily be guided by shooing. They do not flock together, there is nothing on their bodies to hold on to, and a mature sow is far heavier and stronger than most of us. They have a mind of their own, and will walk through you if you happen to be in their way. The only thing that stops them is a wall, and despite their undoubted intelligence, they are easily persuaded that a board is a wall.

The art of the exercise is to hold the board – usually about three feet wide, two foot six deep – low and just guide the pig's head with it. Without too much huffing and puffing, she will usually end up where you want her to go. So it is with this companionable sow; we find an empty, well-bedded arc, and shut her in it for the night. David will let her out first thing tomorrow, but at least she can't just scarper back to her friend now. And I can see that getting the dividing wires erected is going up David's priority list.

*

Next morning I venture down to Lower Farm again. I've not spent much time here since we got back, and the place is being transformed almost daily. This 210-acre farm became mine a couple of years ago; it's the first time I've owned any land myself, and this allows me to do things that my landlords would be less keen on at Eastbrook. These 'things' mostly involve trees. As I mentioned earlier, I've been long-convinced that trees and other perennial crops should be a much bigger part of our future. As farmers, we set much store by growing annual crops like cereals, oilseeds and pulses, all of which need to be planted each year. This means ploughing, usually, or spraying before direct seeding, then cultivating the soil, planting and rolling and weeding (or spraying some more instead, if you are that way inclined). It's a lot of work, even before you get to the harvesting and storage process, and disruptive to our soils to boot.

Growing grass and clover helps with all these problems, and will build soil well if managed correctly. But we don't eat grass, so this means you need to keep animals. Fine by me; I would be very happy to only farm livestock, but man probably shouldn't live by meat and dairy alone! As was very apparent at the Oxford conferences, the debates about meat-eating and its impact on climate change mean that we should be moving towards eating less of it, so how do we grow other foods that are good for us, and better for the environment too? I am sure it will be many moons before we stop growing cereals, our staple food in the West, but tree crops could be a much bigger part of the picture.

One of the many great things about trees is that they grow upright, catching sunlight from a different angle. As farming is essentially the art of catching sunlight and turning it into food, this means that an acre of land can produce much more if it has trees and shrubs as well as a flat crop. It's three-dimensional farming, if you like.

Trees also cut the wind, and hold soils in place. Animals are always happier when they have trees to shelter under, both in winter and in summer. I see how my pigs will escape to forage in the hedgerows if there are some close to their paddocks. Sadly, on our wonderful open downland, there is not much of this terrain for them. We have to provide artificial shades to protect them from heat in the summer, and wallows too. They would love some trees!

Trees can also help nature thrive, providing food, shelter and nesting sites. They can draw nutrients up from deep in the soil, below the level that most crop roots will penetrate, so that minerals that might otherwise leach into watercourses can be captured, and some of them recycled into the topsoil as leaf litter. And they store carbon, vast amounts of it – if not as much as the soil itself – so at a time when we desperately need to reduce carbon in the atmosphere, trees have got a big role to play.

Until recently, we've tended to think that trees live in woods or forests, and, of course, that's a good place for them. But we've divided our land into 'woodland' and 'cropland' without thinking of the benefits of mixing them up. That's what I'm interested in: growing productive trees, whether for fruit, nuts, timber or biomass, on land which will also be cropped or grazed. This approach is called agro-forestry.

Of course, we have always had orchards, sometimes (but rarely commercially) with animals grazing within them. In the UK, though, we've destroyed most of our orchards, and now import nearly 90% of our fruit. Growing fruit without chemicals is not easy in England; our warm, wet summers mean that diseases spread readily. In a traditional orchard, with trees close together, often on rootstocks that are designed for yield and ease of harvest rather than resilience, pesticides feel like an essential tool to many growers. But if we spaced the trees out more, this should be much less of a problem.

So what we are doing at Lower Farm is a giant, very long-term experiment, albeit one we have some experts guiding us on. We are starting to plant the fields with trees in rows that are 27 metres apart; this allows us to crop between them if we want to, and if the farm stays in grass, we will have sensible-sized paddocks for grazing.

It started with an overall farm design, which was led by an inspiring young woman called Amelia Lake who specializes in agro-forestry and permaculture systems, and worked with us for several months, along with Ben Raskin, our Head of Horticulture at the Soil Association. As with all the best laid plans, however, you need to evolve them as you learn more, and that's certainly happening here. The first field we planted was Barn Field, a small (eight acres) sheltered patch close to the farm entrance. This is the most intensive and varied mixture, where we are testing out a number of tree crops on narrower 10-metre-row spacings, and planting soft fruits and sea buckthorn, which is increasingly thought to have powerful anti-oxidant properties, underneath the trees too.

There are many deer at Lower Farm, so we have fenced the

field with a high mesh to stop them getting in, and with rabbit netting too. Nothing will stop the squirrels, however, and we still need to work out what to do about them. It makes a perfect spot for chickens: safe from foxes, and with all the trees to forage under. We had a house of laying hens here last year, and were surprised at how well they kept the base of the trees weed free, once we had mulched with woodchip. Weed control is one of the greatest challenges in establishing trees organically. In virtually all new non-organic woodlands, vegetation around the trees will normally be sprayed for the first few years with glyphosate, the controversial herbicide, which, of course, is not permitted under organic standards.

But despite having created what seemed to be an ideal habitat for chickens, we had not anticipated the aerial threat. We started finding dead chickens in late summer, and were mystified. A number of birds of prey spent time around the field, attracted, we assumed, by the multitude of voles that have proliferated along the slots into which we plant the trees. Voles can do damage to the trees, and when planting woodland, we often put in high perches to allow owls and other birds of prey to hang about comfortably while waiting for the voles to break cover.

But it was the red kites rather than other birds of prey who seemed to be spending the most time there. I started to watch the field from the 'hide' of my truck and finally saw one of these magnificent birds swoop down into the longer grass on the edge of the field. Dashing across I found a still-warm chicken carcass. I didn't actually see the kill, and our conservation expert, Clive, is still adamant that red kites' feet are too weak to grip such a large victim. I'm not convinced. Either way, we will need to remove the chickens unless we find a solution to protecting them while the trees mature and provide more cover. Meanwhile, we will also have to find other ways of keeping the weeds under control.

As well as planting the agro-forestry trees, we have started an ambitious project to establish more woodland. This will help protect the more fragile productive trees in the fields from the prevailing wind, and in a decade or so, once the trees are well established, I hope to be able to experiment with introducing livestock – cattle, sheep and possibly pigs – into these pockets of woodland for short periods, so that they can browse on a wider range of plants, and benefit from the shelter.

A chalk stream runs through the farm. It is the same stream that arises at the watercress beds, flows through the village pond, then on down past the church and pub and away on to the clay land. It is overhung by many willows, which shade the water and have started to split, so this winter we have pollarded or coppiced most of them, to allow some light to the stream. Looking around now, branches lie in huge heaps, awaiting chipping. Some will be good enough for firewood, once they are well seasoned, but most will be used as a mulch for the next generation of trees that we are planting. Against the stream we have planted a long strip of native woodland; it will make a glorious walking route for us in our dotage.

By the close of day, the weather has turned. The wind whips across the pig field, and the little ones are reluctant to come away from the shelter of their arcs. Their paddock is a sea of mud, and it's hard to traverse without slipping. It seems this morning was a false dawn; will this winter ever end?

10

The Drugs Don't Work

21 February

The piglets are eight weeks old now, and under normal circumstances they would be weaned this week. But this is not a normal winter. With staff sickness and tough weather conditions, work is still running behind schedule. And on the horizon is the talk of snow and high winds, the 'Beast from the East' heading our way. The piglets may be better with their mums if the storm is as bad as the forecasters are predicting.

Of course, on most farms the piglets would have been weaned weeks ago. As I mentioned previously, three to four weeks is the normal time on more intensive farms than this one, so that the sow will get back into pig again quickly, and maintain her 2.4 litters per year. While she is lactating heavily, she will not come into estrous and become fertile; this will only start again a few days after the piglets are weaned. Lactation is not a reliable means of contraception, but with pigs it is only when they stop producing large quantities of milk that they will come into season.

With early weaning comes the risk of digestive upsets and infections, as the piglet's gut is not mature enough to deal with some types of solid food, and without the protection of the antibodies in her mother's milk there is the challenge of bugs like *E. coli*. The stress of weaning – being moved into a new environment – doesn't help either. Many piglets on normal, non-organic farms will have zinc oxide and possibly antibiotics added to their feed to help prevent or stop an outbreak of scours.

With all the concerns about over-use of antibiotics in farm animals, and the evidence that this can lead to them being less effective in human medicine, the pig industry is trying hard to cut down on the quantities given to young animals. Not long ago, most would have received several doses in the weeks after weaning. Even at Eastbrook, when we first started keeping pigs, we would need to treat individual piglets, often before weaning. For us, the breakthrough came when we started using a clean paddock for every farrowing, rather than using the same one twice. This stopped the problem overnight: cleanliness is next to godliness it seems.

The Soil Association has been raising awareness and campaigning on this issue for over 20 years, much to the annoyance of those whose intensive systems rely on these drugs. From the time that antibiotics were first discovered, it has been known that the bacteria they destroy can become immune to their effect. Every time bacteria are exposed to an antibiotic, a few will mutate to develop resistance; the more an antibiotic is used, the more likely that resistant populations will be selected for and grow.

Finally, the message is getting through; the Chief Medical Officer has talked about 'a dreadful post-antibiotic apocalypse' which will mean that before long even routine surgery will become deeply hazardous. It is predicted that by 2050 ten million people worldwide will die needlessly each year because of antibiotic resistance, so this will be a bigger killer than cancer. Of course, most of this is down to over-prescribing for humans, but animals are given nearly half of the antibiotics used in the UK, and for some types, like those effective against *Salmonella* and *Campylobacter*, it's the over-use in animals that is the prime cause of resistance. The resistant bacteria can then be transmitted to people, through being in food, or directly from animals themselves. Pig farmers in Denmark are routinely screened for the antibiotic-resistant bacteria MRSA when they go into hospital, as it has been shown that they are much more likely to be carriers than the general population.

Many farmers are now trying to reduce use, but this can be hard without a fundamental change of system. Improved hygiene can help in intensive units, and Denmark and the Netherlands have reduced antibiotic use significantly, but it is only when changes like later weaning and giving much more space are made, as they have done in Sweden, that the levels seem to fall very low on a national level.

It's important for our animals too, that antibiotic resistance is avoided. When we do need to use one, we want it to work. While we never have to treat animals for scours these days, we occasionally get a lame pig or a case of metritis, and once we had a batch of poor, low-immunity piglets who got a skin infection.

But as long as we get our husbandry right it's fairly unusual that we need them. The science shows that many of the things we do, like weaning late, not changing their diets dramatically at weaning – we make sure that they stay on the same feed they have been eating while with their mothers, so that they only have the loss of milk to cope with after weaning – reducing stress and frequently moving the pigs on to clean ground, are proven to help avoid the need for drugs. I like the title of an article in the science journal *Nature*, 'Dirty Pigs are Happy Pigs', showing how the gut flora of outdoor sows had oodles more of the healthy *Lactobacillus* bacteria. Eat more soil, I say!

But while we are making progress in the UK, and in some other EU countries, that is not the case worldwide. There have been scandals about the profligate use of Colistin, an antibiotic drug of last resort for people, on pig farms in China, and a recent report from the Soil Association tells us that the USA uses five times as much antibiotic in pig farming – and over 16 times more in beef production – than we do in the UK. Another good reason to hope that we are not flooded with American meat products in the bold new trade agreements that some in government seem so keen on.

At the end of the week, Molly and her mates finally get their first vaccinations. They are overdue, but better late than never. In these wet conditions, even this task is a trial. On Thursday evening, David sets up a pen made from metal hurdles in the paddock, beds it with straw, and then removes the feeder. The only way to catch our free-ranging pigs is to lure them with food, therefore they will need to be hungry the next morning so they will follow him anywhere for a bag of feed. And it works, they trot happily into the enclosure – in fact, some of them are already there, enjoying the straw and seemingly not bothered about the lack of a roof. Each piglet is caught, squealing as though it were being murdered, but it's a momentary job, one quick injection from the flagon around David's neck, and a spray mark so that he can see who has been treated. Once they are all done, the hurdles are opened up, and the feeder reinstated. He spreads a few buckets of feed on the ground too, so that the weaker ones get an opportunity to fill up quickly.

I ask him to pop in to do our little orphans too. They are still struggling, despite our endeavours with hot-water bottles, whey and titbits. I'm worried about the way that they are constantly scratching; all pigs love a scratch, but there's something going on here that needs attention. Skin conditions, like mange and ringworm, always seem to set in on animals with low immunity, and while there's no sign of a problem in the pigs in the field, our sickly trio need attention.

The three of them have started to develop distinct characteristics, despite their rather forlorn state. It's come as no surprise that the feisty one, the bigger runt, the one that wants attention and eats the most, is a male. He fights the other two at the food bowl. He butts them with his snout as the whey is poured into their bowl. He's the first to emerge from the straw as we open the gate. He's the one more likely to be sleeping with his head towards the gate, as if he wants to know what's going on. He is pig Number 1.

The one we shall call pig Number 3 is a female with the most pronounced runt shape, with a hunched back and bad skin, and would appear by any reckoning to be the least happy and least confident. Her tail has remained resolutely straight. She is completely uninterested in any human contact and is the least engaged when we approach with feed, or whey or clean straw, and mostly moves to the back of the pen whenever we're around them. I assume that over time she will get used to us. I've told myself there's absolutely no reason why she should ever like us, whatever that means to a pig, and if that does matter to us, we should prepare ourselves for some future disappointment!

Pig Number 2 is another boy. He is becoming friendlier, but still much less so than Number 1. He is the piggy in the middle, a sort of hybrid of the other two.

Some mornings we go to see them and there is absolutely no sign of life. It's slightly disturbing. Then we call out to them, 'Morning, pigs,' and the breathing rate increases and the straw starts its low rhythmic rumble, so you can at least see roughly where they are. You look more carefully, and there's a pair of eyes looking out at you through a veil of straw, a little like those crocodile eyes poking out of the water that you see on TV natural history shows or horror films, with the rest of the body hidden. You see that the feed's been eaten, the whey has been drunk, so all is apparently well. Up gets Number 1, Number 2 follows, Number 3 waits. Whey into their bowl: Number 1 is first at it, and when Number 2 tries to muscle in, he gets a repeated butt in the side of the head. So he goes round the other side to try a different access point and to get out of the way of the violence, and the bully just follows him, and does the same again. Number 3 can't even be bothered to join in, and she skulks around the dry pig-feed bowl, and takes some water. It's not warm enough to hang around and very soon they dig back into their fluffy straw mound, and resume their endless sleeping.

*

The weekend brings hard frosts again, and a chance to get silage bales into the growing-pigs' paddocks. At this stage of the winter, there is not a blade of grass remaining in the paddocks where pigs have been for more than a couple of months. They love to eat something other than dry food, and while they will munch on straw when we bed them up, silage is much more tasty. We don't want to make any more mess than we have to, though, and moving the half-tonne bales requires a JCB (other loader tractors are available!), which can churn up the ground even more. So for once, frost is our friend.

The starling murmurations are extraordinary at the moment. On Saturday evening I walk up to Lammy Down, the highest point along these hills, as a tiny group of 50 or so birds sweeps noisily over us. I look up, and much higher there is a vast gathering of maybe 50,000 starlings, perhaps many, many more. Other small flocks fly in, always from Eastbrook, seemingly saying, 'Where's the party?', and then the big cloud starts to break up and fall towards the ground. For 20 minutes, I watch ten or 11 separate murmurations. At moments they are all together, in one vast black cloud, then suddenly breakaways scatter side to side, lower and higher. I have to move my head from left to right, up and down, to keep up with what is happening across the entire skyline, against a stunning winter sunset of blue, yellow, pink, white and grey. I am watching my own astonishing natural wonder, all alone, apart from a bored Dog: starlings tumbling, merging and separating, small flocks cascading downwards out of the bigger ones, falling like unfurling bomblets as they approach the ground, then regaining height as they meet another, larger group. I am reminded of those Slinkies we had as children. Those amazing metal spirals that could walk down the stairs, and the way their metal loops opened and closed as they moved, when gravity took them. And those TV pictures of tornadoes sweeping their way across the American Midwest. These starlings are doing all of that, spiralling, swooping, tightening together so the sky goes black for a moment, then opening up to let the light through again. As the sky darkens, gangs split off from the main event and make a rush for the Lammy Down woods, then having got there, screeching as they fly low over the trees, they decide the sun has not gone down after all so they go back to the party! Around 5.35, some 20 minutes after my personal display had started, the first birds decide it is finally time for bed. Group after group make for the same noisy patch of woodland, while Dog and I, now very cold, head back down the hill.

The forecasters are still warning about the arctic conditions heading our way next week. Meanwhile, we have plenty to distract us from the ominous news. We have been approached by a large organic estate a few miles away, who are keen to see if we would be able to help with their farming. I know the farm a bit already; it's a simply stunning downland expanse, which has been enriched by tree and hedge planting, and the reinstatement of traditional dew ponds since the owners bought it some 20 years ago. Their farm manager is leaving, and they need to find a way forward that does not entail them having to get too involved in the day-to-day arrangements. Henry, Sophie, Dai and I make a first visit to start to explore the options, touring the farm, checking the cattle, buildings and grain dryer, and trying to work out what might be the best option for them and us. With the new dairy already on the cards, it feels as though this is possibly a stretch too far, but at the same time, it is a perfect fit.

The downside of our dairy expansion plans to increase the number of cows we milk is that we have less room for other things, like arable crops, our beef and sheep, and, of course, the pigs. While Sophie and Dai's passion is for dairy farming, and it makes sense financially too, we will be producing many more calves, and be unable to rear them all on our acreage. I've always felt that we should ideally be able to take responsibility for the calves we breed, and make sure they have a good life. We have found someone to buy them and, indeed, some of them will go to my sister's farm, but I would prefer to keep them if we can. Even more, though, I've been anxious about the future of the pigs. As they rotate around the land that is closest to Starveall and the new dairy, we have a plan to reduce the herd, but this will leave our marketing business short of pigs at a time when few other farmers seem keen to start organic herds.

So this farm might be the answer to many of my concerns. It is light land, ideal for pigs – although possibly a bit flinty, we would need to watch out for lameness – and there is enough room for all our beef cattle and sheep too. But it is a huge new venture, and we will need to think very carefully before we plunge in.

While we are exploring the buildings on the farm, I get a text from Tim. 'Help! Big problem with one of our house pigs. Do you know any vets?!' I'm standing next to Sophie, who gets on the phone to him, and we dash back to Eastbrook.

Tim's had to leave to run the pub by the time we get there, but the problem is obvious. Number 1 pig has had a rectal prolapse, something I've never had to deal with before. He seems remarkably unconcerned that part of his inside is now outside, and is only distressed when we try to separate him from the others, and when Sophie attempts to rectify matters. Unsuccessfully. This is heart-breaking. A jolly little pig, seemingly oblivious to the rupture, but who will surely start to be very uncomfortable soon. His panic at being separated from the others makes it doubly difficult to deal with. There is a decision to be made, and I try to call Tim; he has been their dad for the last few weeks, caring for them night and day. I can't get a response. This is all happening so fast, and I know we have to make a decision. If he would be happy separated from his friends, then it might be possible to try to treat him further – but he panics every time we attempt this. I cannot bear it, yet there is only one option it seems under the circumstances. We use a humane killer, a captive bolt, to euthanize small animals; he is so tame and friendly that at least he is completely unaware.

This was Number 1 pig, the one who was always first to feed, the one always keen for attention, the dominant bully, the one who had bonded with us. The other two seem very subdued. They saw nothing of the end of their sibling, but they hide away and are reluctant to feed.

I dread telling Tim. He will take this hard, even if he doesn't show it.

11

The Beast from the East

The farm team starts to prepare for the storm that is due midweek. The wind is expected to change direction, from the prevailing south-west to a north-easterly. All of the arcs have their doors facing that way, and I am worried that snow may be driven into them. Newborn piglets would not stand a chance. I ask David if he can get bales in front of at least the youngest litters, and those that are due to give birth. By late afternoon there are trailers full of bales arriving, and a JCB carefully positioning them to give as much protection as possible. We still have a few days to prepare, but I'm delighted that they have made a start.

Molly's gang have new neighbours in the guise of a bunch of weaned pigs who are gallivanting around the grassy paddock next to our now very muddy one. I sometimes wonder whether the unweaned piglets look over the fence with the wistful gaze of ten-year-olds who long to be part of the teenage world. The teenagers are endlessly energetic and exploring, clubbing together at the slightest provocation in mock hysteria, sprinting around the paddock before coming back for another look at whatever it was that set them off.

The vaccination pen is still in their paddock – a sign that storm preparations have taken priority over routine jobs – and Mrs Messy Bed seems to have taken up residence in it. Always a loner, she seems to prefer to sleep under the stars, at least when it's not raining. This is her fourth litter, and she's done a reasonable job so far, which is rather surprising as she seems quite happy to abandon this lot for a life of solitude and contemplation at any opportunity.

Molly's mum is one of the four gilts in this group. All were born in early March last year, so they have had their first litters a little earlier than would normally be the case; we would expect gilts to farrow for the first time at just under a year of age. Maybe that's why they have stuck together so much. They have been reared as a group, and there's clearly still quite a bond, especially between the two that refused to be separated at farrowing, 312 and 314. The one with the Duroc genes, 302, is rather smaller than the others, but chunky and very assertive, and usually the one who is most keen for some petting. She is the most playful too, bucking and weaving with joyful energy when she has any excuse. Molly's mum is 309, more delicately featured with narrow shoulders, and she is more high-handed in her manner too; she hangs back from the melee that often takes place when I visit, when swarms of piglets and their mums almost knock me down in their enthusiasm for scratching and boot nibbling.

The eight piglets that remain from Tess's litter are still notably smaller than the rest of the group, but much bigger than our house orphans. Yet again, I wonder if we did the right thing removing the orphans from their gang. The rest look bonny, bar the two Saddleback-like piglets, which are seemingly under the weather, and I'm not sure why. They don't appear unwell, just poor doers. I am sure that they are Messy Bed's and I wonder if the communal approach has let her off the hook on maternal duties and she just expects the others to feed them! I wonder about taking them home too, but given that our orphans are still not thriving – and we have lost one of them already – I think they may be better off with their family group for now.

*

On Saturday night the temperature starts to dip. It is ferociously cold, though no snow yet. We start supplying hot-water bottles to our pair of orphans again, very aware that there are now only the two of them to keep each other warm. Whey and water freezes so quickly, and needs replacing several times a day. They have rather gone off the whey, it seems, and for a few days get excited by warm milk. However, that's a short-lived fad too. I try an egg. That appears to get their interest and stimulates them to eat some of their proper pig food too. But mostly they stay tucked up in bed under a mountain of straw.

Sunday is brilliantly cold, crisp and clear. There have been several new farrowings overnight, and many of the week-old piglets are out and about on the frost, looking fantastic. But when I see David, he tells me that he has had another two prolapses; this is very worrying as we have never had a problem with this before. When I get home for breakfast I do some research, and ring our vet for a chat. It seems that cold can increase the risk, as piglets sleep in heaps for warmth, and this can create too much abdominal pressure. Gorging on food can be an issue too, as well as over-rapid growth, and sometimes diarrhoea – we haven't seen any sign of that. Given that this is the coldest spell we have seen for many years, and that the prolapses have been in groups that are multi-suckling so that there are more piglets together, I suspect it is caused by the huddling they are doing for warmth: even more straw seems to be the only solution.

It's cold in the farmhouse too, and I go to bed with my own hot-water bottle. Even with that, and an enormous duvet, I'm shivering overnight, and anxious about the two in the dog kennel as the thermometer drops below minus eight.

I wake, still shivering, well before dawn. I stumble out to check the pigs. Tim's away, so I'm feeling the weight of responsibility for the remaining two! But I'm reassured by their greeting, even at this unearthly hour, and the fact that their hot-water bottle is still warmish. I give them a little warm milk and another egg. Yes, it's definitely the egg that wakes them up. I think we are on to something here.

In the field, Dai is helping to get water round. This is crucial and even more important than feed. The trick is to get the volumes right so that the troughs are empty by morning, otherwise they are a solid block of ice and you can't get any more in. That means a thankless task of sledgehammering out the ice, with each trough taking 20 minutes of sweat and over 100 troughs in the field.

There are bales around most of the farrowing arcs now, but the majority of the growing pigs still have exposed doorways. It seems a Herculean task to protect them all, and I am sure that these older pigs will be less vulnerable. Even so, I would be happier if they had a windbreak too.

Messy Bed is still lying out in the vaccinating pen, despite the below-zero temperatures. And when I go to see the litters, I find that one of the poor Saddlebacks has died, just lying amongst the group. Maybe I should have taken him home too, when I spotted him a few days ago – but he did have a mum, just a rather self-indulgent one. It's unusual for a sow to be such a negligent mother. I've had a bit of a soft spot for Messy Bed, with her independent streak, but I'm cross with her now. She could have made a bit more effort.

Tuesday and Wednesday are bitterly cold too, and the only upside to this is that Number 2 pig, the boar, becomes increasingly pleased to see us. At least, he is starting to associate us with good things to eat, and is even prepared to tolerate the odd bit of physical affection as part of the deal. There's a fascinating dynamic between the two siblings: they both like eggs more than any other food, and the gilt will shunt him away from the bowl when there's egg at stake. He gets his own back when it's dry food time. Even though there's plenty of room for them both to feed, and no shortage of it either, he refuses to let her in.

*

I need to leave early for London on Wednesday, the day the storm is due to hit. It's hard to know how bad it will be today, and whether to abandon ship, but I have a full day of meetings planned, including with a senior adviser to the Secretary of State, so I'm keen to make it if I can. Although Michael Gove and his team are making all the right noises about prioritizing nature and farm-animal welfare in future agricultural policy, they are seemingly coy about mentioning organic farming as one easy way to help deliver these benefits. It would be very helpful to understand why they are so reticent, and to have the chance to make the case more strongly. Likewise with soils. Gove has been clear that protecting and reinvigorating our soils is a top priority; scientists believe that we may have only 60 harvests left if we keep eroding them as we do currently. Without healthy soils, humanity cannot survive! But the most important indicator of soil health, the levels of organic matter it contains, has not been included in the new requirement for farmers to test their soils regularly. It seems like a huge missed opportunity to build a picture of how our soils are faring, and to encourage farmers to concentrate more on improving their quality in this vital regard.

I get to London to make my case. When Laura, our Head of Public Affairs, and I arrive at the department's offices, we are told by the MP's assistant that we have only 35 minutes, as there's a vote in the House that she must attend. It's not much time to set out the facts, but we do our best – and to be fair she has a good sense of the importance of soils, just less of a practical understanding of what to do about the crisis.

'The key thing is to get farmers to test and understand their soils,' I emphasize. 'And it's important that we do this ourselves, rather than just leave it to the experts – even though the results will need experts to check them once in a while, especially if government is paying for them.'

She quizzes us about what should be measured, and how you can tell if a soil is healthy.

'Probably the most important thing is soil organic matter,' I explain, 'and the microbial life it contains. While we don't know much yet about all the billions of bugs in soil – some of them haven't even been identified, let alone what they do – we do know that earthworms are a great indicator of a healthy soil. So getting farmers to count them at the same time each year, ideally in the spring before the ground dries out and the worms burrow deep down, would help us appreciate how our farming methods can be improved to enable worms to thrive. Anything that gets farmers to put a spade into the ground regularly, to get to know what good structure looks like, and how the earth smells and fractures when handled.'

'But surely farmers have every reason to care for their soils, if they are so important to maintaining yields?' interjects the MP.

'I agree, but fertilizers and pesticides have masked the problems the soil is facing for decades, and many of the problems we are now aware of have been building up for ages; they will take a concerted effort to reverse. And if farmers have only short-term tenancies on their land, or worse still, are just using it for a year to grow a specialist crop like potatoes, they may not have the incentive to think about the long-term impacts of their management. Farmers are so squeezed financially that they often feel that they cannot consider the long term, just this year's profits.'

Laura chips in and explains that we would like to see a soil database, generated from farmers and soil experts too, which would allow a clearer picture to develop of the state of our soils, and what is working best to deepen and reinvigorate them. Government could facilitate this, and require soil testing and data submission as part of future support arrangements, so that there is an incentive for farmers to examine their soils and send in the results.

The MP's assistant is looking at his watch, and gently suggests that it's time to leave. I'm very keen, though, to find out why there is so little enthusiasm in government for organic farming as a system that is delivering soil health and much more besides. 'I'm not sure,' the MP responds, 'but I will try to find out.'

As she starts to gather her papers together, I suggest she and the minister might visit Eastbrook, to see the techniques we use first hand. Her assistant is now agitated; our time is clearly up, so we part warmly, and I leave in the hope that this brief meeting has sown a seed or two. You never can tell.

*

As I traverse London later on my way home the storm is picking up all the time, dark grey skies and snow blowing about the streets. It feels deeply ominous, like the beginning of a disaster movie, with the wind whipping through the rapidly emptying urban landscape. Thankfully, the trains keep running and I make it back to the farm.

The next morning is a different matter. I wake to three inches of snow, and a howling wind. The radio tells me that roads are blocked and many trains cancelled. I am supposed to be in Bristol, but it's clear that I'm not going anywhere.

As soon as it's light, I set off for the pig field on foot, wrapped in a dozen layers to protect against the blizzard and with a rather reluctant Dog in tow. I want to see whether the road to the Downs is passable, so I head up across the Nells – the fields in front of the house – cutting back towards the road just short of the silage clamp. The snow is drifting to the extent that it is not possible to go further across the fields anyway; the fence lines have trapped the fine mobile snow so that it is three feet or more deep against them. The road is worse still. Running parallel to the direction of the wind, snow drifts have accumulated wherever there is a break in the hedge. The strange eddying shapes are rather beautiful, but the road is blocked to me and even to tractors. There's some work to be done before anyone will reach the pig field this morning.

I retrace my footsteps back towards the farmyard. It is still only 7 a.m., and I may as well go to report in to Henry and the team, who will be meeting up shortly. I have warmed up now, but as we turn back, Dog decides that he has had enough of this stinging snow, and belts for home. I find him sheltering under the wild damson bushes on the Valley track, and we head off to the yard. Henry organizes a makeshift snowplough in the guise of a JCB with a bucket on front and the team spend the first couple of hours of the day clearing the road. The cows must be fed, however, so the feeder wagon trundles across the frozen fields to get to the silage clamps. By mid-morning, David makes it to the pig field to do what he can under these appalling conditions.

One of the benefits of our investigation of the local estate is that we have met a young woman, Rachel, who has been looking after their small pig herd, but was in need of a new job now that the estate could no longer keep her on. Dai has been in touch with her, and today is her first day on trial with us. What a baptism of fire! Impressively, she makes it in, and soldiers on with David in the storm. It is a day to get just the basics done, and everyone retreats to the pub for lunch and a defrost. I am very keen to get to the field and see for myself how the pigs are coping. But it is not possible on foot, or even in my truck, so I have to rely on their assurances that it's not a disaster zone.

The house pigs seem OK, though very wary of the snow. I ferry hot-water bottles and warm drinking water, and pile even more straw into their kennel. Working from my kitchen, I keep an ear on the news. Apart from the travel chaos, it seems that shops are running out of food, causing extreme crossness from some people who feel aggrieved that with one short storm, daily essentials are becoming unavailable. While we fret at home about two small pigs, and David and the other farm staff labour manfully, and womanfully, against the elements to ensure the best possible welfare for their thousands of pigs and hundreds of cattle, the country's millions of shoppers have been stripping shelves bare in their preparation for a long winter siege. To compound the problem, the food supply industry is also very rapidly grinding to a halt.

In the sector we know best, the abattoir and meat-processing industry, staff are unable to get to work, water pipes and machinery are freezing, lorries of live animals and whole carcasses are stuck in snowdrifts. Slaughter is delayed or cancelled, meat cutting the same; stuff isn't getting into the factories, and so nothing is coming out of them. Demand has surged, and supplies are now almost non-existent.

There's been a lot of energy and thought gone into ensuring that everyone, everywhere, gets their food delivered 'just in time'. It's a management theory put into practice in this industry and most others to help businesses hold minimum stock – holding stock is expensive and requires people and space to manage – but still always have enough on the day it's needed. The advantage of 'just in time' in the food world is that, in theory, everything on the shelf is as close to fresh as possible, even though there's plenty of playing around with use-by dates it seems – but along comes a mini natural disaster, such as a very cold spell of weather, and it all falls apart.

One of the major reasons for this is that the vast majority of food, probably the most important part of our daily lives, is no longer a local business, but a complex web of national and international trading. We're almost as guilty as anybody, in that while our animals are slaughtered quite locally, some 30 miles away, the cutting, processing and packing takes place all over the country and, indeed, in Germany too. It's near impossible to find businesses that will manufacture the smallish batches we need, especially in organically certified facilities. For most farmers, processors and shoppers, food miles are a huge problem. The food retailers need deliveries seven days a week, every week of the year. The only time we are not required to deliver our organic bacon and sausages to our major national retailer is Christmas Day and Boxing Day, which seems crazy when we deliver products with shelf lives of more than 20 days. Even in this small country, some livestock travel hundreds of miles for slaughter, and milk can be collected from farms in Wales and processed in factories in Scotland. It seems madness that we've allowed this to happen in our drive for efficiency.

Where we as a business are better at localness is in our milk processing. Our on-farm mozzarella factory is still running, despite the snow, although the staff look a bit chilly and steam is shooting out of every orifice the building offers; and the rest of our milk is on its way to the Goslings' Berkeley Farm dairy, five miles from here. Our vegetable supplies for the pub and Chop House come straight from the fields at Pete Richardson's Westmill business, a superb organic smallholding four miles away. Almost every day, he digs his winter greens for us, and delivers them, along with anything else we want that he is able to get out of the ground or retrieve from his straw-based storage system. Today, we find ourselves with more swedes and root crops than usual, so we start making Cornish pasties for our intrepid guests. Localness is having some benefits beyond the feel-good factor in these unusually harsh conditions.

Perhaps this will be appreciated more widely now, though I know how short our collective memory is. Government and local authorities could start to encourage more local supply chains and the infrastructure they require, such as abattoirs and processing facilities, so that everyday essentials like milk, bread and vegetables become more available locally. The benefits could be enormous: to the local economy, to farmers and consumers, and to farm animals too. People could eat fresher food, which would retain more nutrients, and require less packaging, and would still be available when a snowstorm or other minor catastrophe hits. Farmers should receive a more appropriate slice of the retail value, allowing them to farm wisely, without cutting corners. Animals would be slaughtered closer to the farm, so they were saved the stress of long-distance travel. And the local economy would thrive with all these transactions taking place within the region, rather than being siphoned off into faraway places. What a wonderful world it would be!

*

Overnight, the wind is wicked. I wake at 4 a.m., worrying about the orphans, and even more about the whole herd in that exposed field. Creeping out in the dark with warm milk, I'm reassured that the orphans are still relatively snug, albeit that their outside run is under a foot of snow. The little boy gobbles up the milk, but the girl is not interested. She used to like the whey, and I can hear the Italians on early-morning cheese duties so I pop round to collect some. 'Warm whey, yum!' she says.

I try again to get to the field once it's light, but again it's impossible. Finally, late in the day, the wind drops a little, and Tim and I set off to see if we can make it before dark. It's a slog through drifts and spitting snow, but we warm up and almost enjoy the adversity. We arrive just as the light is starting to fade.

It is an extraordinary sight. Every arc has a large bale or two of straw in front of it, and every bale has a snowdrift behind it. Over 120 arcs, and only one has any amount of snow inside, where the bale was at slightly the wrong angle. (The inhabitants of that one had quite understandably decamped elsewhere.) In the luminous white light, pigs are everywhere, rooting and feeding. Molly and her mates gather round on the straw in front of their arcs. Some are snow-eating; I'm not sure if this means they are thirsty, or whether it's a novelty for them – ice cream on tap. Their tails are rather straight though, and I guess they haven't fed much outside during the peak of the storm, but they are all alive and not seemingly much the worse for wear.

By the next morning, the snow is melting. After the hard frost, there are leaking water pipes, even while the guys are still bowsering. It seems especially unfair to have both these jobs to do at once. The pigs seem all right, but their dung is a bit loose. I wonder if it's down to all that snow-eating, along with intermittent feeding. The slushy snow is starting to make their bedding damper than I would like, which is hardly surprising when they are traipsing in and out with the white stuff on their feet. They will need more straw later if we can possibly get it there, alongside all the other tasks the team need to do.

By Sunday most of the snow has gone, though some tracks are still blocked with drifts, and most of the water pipes remain frozen. There are gusts of freezing rain through the day, and then in early evening, the most stunning double rainbow that I have ever seen, arching over the Valley. Braving the louring sky, Tim and I walk to the Ridgeway, then along to the pigs. The guys have done a fantastic job in bedding-up the arcs, and they have cut some of the protective bales so that the pigs have dry verandas to lounge on again. All of the pigs look snug, and tails are starting to curl. We watch a barn owl hunting along the track in the twilight, and I feel a deep pride in what David and the farm team have achieved this week, heroic in these appalling conditions. Out of over 1,500 pigs, many of them tiny, we have lost just one in this terrible storm. A remarkable achievement.

12

A Minister Calls

19–23 March

With the storm gone, this week should allow us to start catching up with jobs that have fallen by the wayside, including weaning. Molly and co. are now nearly 11 weeks old, and they are next in line to start their new lives away from the sows. As I visit them midweek, with a small film crew who are making a video about the Soil Association for the website of our ethical bank, Triodos, the pigs swarm all over me, oblivious to the imminent changes. It's a raw, cold day, and ensconced in my oversized muddy coat, I'm not exactly camera ready, but the pigs are the stars of the show.

I take the crew to see the agro-forestry, and the newborn calves too. We are not due to start calving for another week or so, but there are always some early arrivals. This morning there are two dainty twins; twins are often born a little prematurely. The start of calving means the start of weaning for the foster cows who have been rearing the autumn-born calves all winter. Even though their charges are now six months old, they object to being separated. Surprisingly, it seems worse than when the calves are newly born, though by now, they are big, strong cattle, and so my nights are disturbed by their calling from the barn next to the farmhouse. The nurse cows are now being introduced to the baby calves, and seem perfectly happy to accept their new responsibilities.

Sophie rings me from Ireland, where she and Dai are looking for young heifers to populate the Starveall dairy next year. We want to bring them on to the farm now so that they can have a whole year to adjust to life in Wiltshire, and as we cannot find organic heifers anywhere in the numbers we need. It also means that they will have undergone a full conversion to organic status by the time they calve. They won't ever be organic themselves, and their meat will have to go into a normal market at the end of their lives, but their milk and calves will be. It's a bit of a leap of faith to be buying the heifers when we haven't finally pressed the 'go' button on the dairy build, but if we don't get them now, there may not be another chance. They have found around 120 that they would like to buy. It will be a big expense, but I agree that they should tender for them.

Our house pigs are still scratching too much, and with their skin looking inflamed in places, I fear they have an infection. There is a condition that I have seen before, many years ago in a poorly group, which is known as 'greasy pig'. It is a *Staphylococcus* infection, and can make them very unwell. I am sure that some parasite, possibly mange, is the root cause, and order a treatment for this. In the meantime, though, I think they need antibiotic cover. As you'll probably have realized by now, I hate using antibiotics. But welfare must come first, and this is a good example of why we need to be able to use drugs like these on occasion, even on organic farms.

*

Back from their cattle-buying trip, Sophie and Dai arrive to treat the pigs with a long-acting tetracycline. They wade into the pen, and the pigs head to the furthest corner. Dai catches the nearest one, the female, and it screeches. He hands it over to Sophie, who holds it tight to her body. It wails, it kicks, it howls, it wriggles. Sophie is completely unruffled. She's obviously handled far more difficult creatures than this one. She has the needle to hand and in it goes, just in front of the shoulder. The injection appears to cause less distress than the being caught and handled. On release, the piglet scurries off to the darkest corner and watches, warily. By this stage, the boy is also in Dai's hands, and it goes through the same routine – noise, fight, anything to get away from these interfering humans. In with the needle, down goes the piglet, off to join its sister in the dark corner. We retreat, leaving new feed and new straw, to allow them to settle down again.

Next day, David pops by to give them their skin parasite treatment. Within 24 hours the scratching has reduced, but by now my concern has shifted to Dog. On our way homewards after a long morning walk over the Downs, he suddenly stops, as though he can't see where he is going. He is staring forward, and cannot seem to move his legs – and when he does, he falls. He manages to get to his feet, but even with his legs splayed wide, bambi-like, he is unsteady, and clearly does not trust himself to move. As I try to help him, talking to him and trying to soothe him, it's clear that he is in another world, barely aware of me and his surroundings. I fear that he is having a stroke. My last dear friend ended this way; tearing around the garden in the mad circles that dogs love to run, she collapsed and never recovered.

I am trying to work out what to do for the best. We are still a mile from the truck, and I will struggle to carry him that far. Gradually he seems to recover some of his senses, so I slip his lead on, and start to guide him home a few steps at a time, talking to him all the while in the hope that something is getting through the electrical storm that seems to be taking place in his head. By the time we at last get to the truck, he has recovered somewhat, but is still dazed and confused. Tim and I rush to the vet with him, fearing the worst, given our previous similar experience. The vet checks Dog's heart, which seems normal, takes blood tests and tells us to keep him quiet until the results arrive, and to video him if it happens again. We are hugely relieved to be taking Dog home with us, albeit apprehensive about what the next few days will bring. My morning walks won't be the same without him.

*

The end of the week sees us all in the office with our advisers, trying to make progress on the three major farm projects: setting up our farming partnership, developing the new dairy, and trying to work out if it is madness to take on more land. It's a head-scratching day, but we are edging closer to decisions all the time. Towards the end of the day, an email arrives with a welcome request: Michael Gove has asked if he can visit the farm next Friday. This is a wonderful opportunity to make the case for organic farming, agro-forestry and local food economies, and I am delighted that he has taken up the offer of a visit that I made some time ago. Maybe his adviser, whom I met on that snowy day in London, has encouraged him to make the trip. I have met him several times in recent months, but always in Westminster, and I am sure that the issues will come to life for him much more when seen in reality. 'Seeing is believing,' as the Soil Association's royal patron, Prince Charles, so often says.

Still absorbing this news, I get moving before the light fades, walking up through the Valley and across Fifty Acres to see if David has managed to wean the pigs today. As I cross the stubbles, I spy some white blobs in the distance, underneath the hedge on the far side of the field. As I get closer, I see that it is a group of mischievous large piglets, escaped from their paddocks and on the rampage. They are having a wonderful time digging in the stubbles and undergrowth, but take flight back towards their paddock as I approach. Quite right, it is nearly bedtime.

My pigs are pleased to see me, still all together in their family group. I am surrounded by a sea of them, and can barely move without falling over. Most of them are so large that I could not lift them if I tried – not that they would let me, but they will let me do almost anything else with them. They seem totally unafraid, and completely recovered from the traumas of last week.

The big straw bales left over from the storm have partly broken up, and the pigs love to lie and play in them. There is quite a commotion in one heap. One of the gilts has wedged herself against the bale, but her piglets want to feed. They are trying to get between her and the bale, squawking madly in frustration.

Then Molly appears, carrying a clod of turf that she seems especially proud of. A couple of the others try to grab it from her, but she holds her head high, and resolutely continues into an arc to do whatever she is going to do with it. Next door, younger piglets play hide and seek, chasing each other through the tunnel between two round bales.

*

Both Dog and our orphans seem to have turned a corner. After a few days of being very subdued, Dog starts eating again, and recovers his energy. The piglets have also started to transform. Their skin looks better, and their appetite is certainly on the up, so we have begun experimenting with some new foods, given their reputation as gourmands. When we tried kale a couple of weeks ago, it was of only minor interest. Nosed at, pushed around a bit, then ignored. But when this week we let them out of their concrete-and-straw kennel for the first time and gave them space around the yard, they made for anything green – weeds growing up from the cracked concrete, grassy knolls sprouting from the seams between the yard and the house walls – and interestingly, they loved the slightly muddy water collecting in the old porcelain plant pot that's been sitting unused in the yard for goodness knows how many years. I've hardly seen them drink the water in their pen, which is always from the kitchen tap. It's our glorious spring water, straight from the chalk, but it's obviously too fresh for them. Needs a bit of mud in it for flavour!

They will fight naughtily over food at any time of the day. Their perfect breakfast is now raw egg, which the female continues to have most of, followed by their ration of organic pig nuts, where her brother is in control. Watching the gilt eat her egg is pure comedy – yellow everywhere, round her mouth, all over her nose, sometimes on top of her nose, sometimes even on her forehead. Then she goes through an elaborate lip-licking routine, before heading for the dry feed – whereupon she is butted hard by her brother, who puts all four feet in the bowl, because he is still in charge when it comes to pig nuts.

All this jollity soon comes to an end. Two weeks after the snowstorm caused such disruption both on the farm and across the country, the chaos recommences with another heavy fall. On Saturday evening I get to the field just as the skies are opening, keen to check that preparations are in place. They seem to be, but my gang look a bit miserable. Unlike the rest of the herd, their bedding is rather damp, and I'm not happy about them facing a cold night without a dry bed. There are some bales open in the neighbouring paddocks, so I trek backwards and forwards with armfuls of straw, putting mounds into each of the three large arcs. The pigs are delighted. The sows immediately start to paw at it, spreading it across the floor of the arcs and saving me much work in the process. The piglets start to eat it, and root through it for goodies. Encouraged, I keep going until it is nearly dark, rather enjoying being useful to them, and warm now from the hard work of carrying the straw across slippery ground.

By morning, there is a blanket of snow several inches deep, more than there was two weeks ago. I manage to get to the field late in the day, and find the pigs marooned on their straw pad outside the arcs, chomping on snow again, but seemingly largely unscathed. Second time around, a more established routine of snow-clearing, and trekking across frozen fields with feeder wagons and water bowsers, kicks in. Again, all normal work is on hold and I wonder whether these piglets will ever get weaned. The farm team are focused on bedding, feeding and watering, and hauling cars out of drifts on the roads around the village. The pub becomes the refuge for many staff at lunchtime, a chance to thaw out and enjoy a hot meal. Our little pigs retreat to their den, occasionally venturing into the yard, but running for cover as soon as another flurry starts. This is all getting rather tedious.

It doesn't last that long though. By Monday night it is melting, and on Tuesday we are shipshape enough to get together with our pig vet, Alex, for our quarterly review of how things are going. We spend a bit of time in the office having a look through the records, which are not as up to date as they should be, due to the weeks of ghastly weather, which have left little time to get the paperwork sorted and on to the computer. Our finished pig weights have dipped, almost certainly due to the conditions. The pigs are using more energy to keep warm, and at the same time, have possibly been reluctant to venture out to feed when the weather conditions have been especially bad.

We start discussing the notion of moving to a three-week batch cycle. At the moment, we farrow, wean and serve every week. Would life be easier for the team if we could concentrate on one of these key tasks each week, so that we would farrow 24 sows at the same time, then the next week concentrate on weaning, and the following one on serving? David is keen to try this approach as he feels it would make his routines more straightforward, one big task each week rather than three smaller ones. My main concern is that with our late weaning, not all sows will reliably come into estrous immediately after separation from their piglets; some of them will often cycle before weaning, as the piglets start to eat solid food, and the sows energy levels rise.

Alex undertakes to do a schedule for a three-week batch system for us to think about further, and she, the pig team, Sophie and I head for the field. As well as checking the health and condition of the pigs, she takes dung samples to check for internal parasites, and monitors lameness levels as part of our welfare outcomes assessment. This is something that the Soil Association has been helping to develop over the last few years. However good the standards are in any system, including organic, it is the management and husbandry that can determine whether the pigs are enjoying life. Overall, she thinks that the pigs look well, especially under the circumstances, but we notice some head shaking, and one or two puffy ears. Is there some mange creeping into the herd? Our two orphans seemed to have it, and they may be the canaries in the mine. She takes some skin swabs and we discuss treatment options in case they come back positive.

Having such a close working relationship with our vets is very important to me. They are a crucial part of our management team in that they help us to anticipate potential problems, and have such experience of what is working well, or not so well, on other farms that they can advise on management and the development of better ways of doing things, as well as on the animal health and welfare aspects. I have often been critical of the fact that vets have been too reliant on drug sales for their income, but that is inevitable if farmers aren't prepared to pay them for their advice. Too often, I hear farmers boast that their vet bills are low. I want my *drugs* bill to be very low, but I am delighted to pay for my vets' time; it is always money well spent with such an excellent vet team as ours, for both the pigs and the cattle enterprises.

We visit each pen, trudging through the remains of the sludgy snow that has left the ground saturated. Many of the paddocks are a sea of mud by this stage of the winter, and the tracks made by the tractors cut deep into the soil. Yet again, we debate the pros and cons of keeping all the pigs outside during the winter, and whether we should develop a more indoor approach for at least the growing and finishing pigs at this time of year. The vets, however, have been cautionary about the idea of bringing the pigs into more confined conditions indoors. Our way of doing things is so very different from other pig farms, and they have been very impressed by the overall health status of our pigs, especially the very low levels of respiratory diseases. Being outdoors with plenty of (very fresh) air, getting lots of exercise and contact with soil, alongside the late weaning, and low stress levels, seems to allow the pigs to stay in tip-top form – just as it does for humans. Overall, they are reluctant for us to change such a winning formula, even though it has a less positive impact on the land – for a short time at least – and on the machinery that has to traverse the rutted tracks, and the people who are working in all weathers too. But the proof of the benefits is such a low level of mortality and morbidity, and hardly any antibiotic use at all.

*

Still chatting away on this perennial topic, while battling the wind and the mud, we inspect every group of pigs, encouraging them out of their arcs to check for any health problems. Eventually we get to Molly's group. There's a marked difference in both them and me. All our pigs are fairly calm with people, sometimes gently curious, but also wary and somewhat detached. But this gang have lost all reserve, and flock around me. I start telling Alex all about them, while scratching and petting them, sitting down with them in the straw, pointing out Mrs Messy Bed, and Mrs Duroc, and the big-bottomed ginger boar, and of course, Molly herself. On this cold, wintry day, hungry and covered in mud, my spirits are lifted by the enthusiasm with which the gang greet me. They push me over in their boisterous playfulness, they sit on me as I lie up against the bales that make such a cosy veranda in front of the arcs, they nuzzle and chew and roll over for tummy rubbing. I'm lost in the joy of them for a moment – and then I see Alex looking rather bemused, and Sophie snorting with laughter at what has become of her pragmatic mother!

Once I've regained my composure, we complete our tour of inspection, and then retreat to the farm office where Alex writes up her notes, and makes some small changes to our health plan. This is an essential part of our Soil Association agreement; we must maintain an up-to-date record of all health and welfare issues, with protocols agreed and signed off by our vets as to how we will aim to resolve them.

After a very welcome break for lunch, we have our monthly board meeting, which brings the senior folk across all of the businesses together to review financial performance and agree any significant changes to our plans. The figures are quite good, except for the pub, which is bearing higher rent and staff costs as a consequence of our smart new rooms – the occupancy of these rooms is not as high as it needs to be. The big farm business concern, though, is the shortage of pigs. Our numbers of finishing pigs on the farm – those that are nearly ready for slaughter – are low due to weather conditions slowing growth, and the other farms we work with seem to be having similar problems. Ideally we would skip a couple of weeks to let the pigs put on more weight, but then there would be no bacon in the stores. Supermarkets get very cross if you don't maintain supplies, so to keep our customers happy we agree to put pigs through that will make little money for the farm.

*

It is the spring equinox, exactly three months since Molly was born, and to mark the day, the first lambs arrive. We lamb outside, allowing the hardy Romney ewes plenty of space to get on with the job with as little disturbance as possible, stocked on grassland that should be sufficient for them to produce plenty of milk. They are lambing in three fields close to the foot of the Downs. Those that are due to have triplets have the best grass; it's hard for a ewe to rear three, so they need plenty of energy. The twin-carrying ewes are in another field, and those that are due to have singles are more heavily stocked. The singles are less likely to mis-mother – a term we use for when a ewe steals another's lambs, which they often try to do when the oxytocin kicks in just before they are going to give birth. As a result, a ewe can end up with a lamb or two that aren't hers, and then her own as well. And they need less feed if they are only milking for one, which is another reason why we can stock them more heavily on the grassland.

*

The government's consultation on the Agriculture Bill is now out, with a deadline for responses by early May. The Agriculture Bill will provide the powers for government to repatriate much of the farming-related legislation that currently resides in Brussels, and this consultation gives us all a chance to give our views on issues like standards and the way farmers may be supported in future. I spend much time with my Soil Association team this week, starting to work up our feedback. While there are many encouraging statements and suggestions in the document, there are still signs of inconsistency over the troublesome issue of free-trade agreements, and whether lower-standard imports will be permitted. This is deeply ominous for farmers and consumers alike.

There will be much work involved in preparing our response to the consultation, and the first thing that strikes me is that, despite the title – 'Health and Harmony' – there is nothing in the document that refers to human health, or even to food. Surely our health, and the provision of sufficient, safe and nutritious food, should be at the heart of future agricultural policy, alongside care for our countryside and wildlife, and for farmed animals too. I am also intrigued by the implications that the question set reveals. While overall, I like the direction of travel, the questions are tightly framed around decisions that have clearly already been made. It's like being asked 'I am going to give you some tea. Would you like Earl Grey or Darjeeling?' when perhaps you would have liked a coffee, or a glass of water.

On Thursday I get back from a couple of days in London and Bristol, to get ready for Michael Gove's visit. I'm looking forward to quizzing him on the omissions in his consultation, and making the case for organic farming as a way of rectifying them. I've been meaning to clear out my rather disgusting truck in honour of his visit, but instead head off to the pigs. After even a few days away, I'm itching to be back with them. The ground is finally drying out, and the pigs are ranging across their entire paddock. They seem extra pleased to see me, and I think I know why. Their feeder has been removed, which means that they will be weaned tomorrow. So they are hungrier than usual, digging ever more avidly for roots and earthworms, and possibly wondering if I bring food. I disappoint them, but they don't seem to hold it against me, rubbing up against my legs and enjoying a scratch. They are looking splendid, even the remaining orphans are a respectable size now, and I feel quite wistful that this is the last time I will be together with them as one big family. It has been a traumatic few months; I could not have picked a more complex group to spend time with, but I feel we have come through some stormy patches together, literally at times.

Friday is here before we know it, and suddenly it's time to stand by our beds for Michael Gove's arrival. I've got a few of the team lined up for the initial chat that we plan to have in the office, before taking him out around the farm. I'm keen that he gets a sense of the breadth of what we do, from farm to plate, and how, with the right help, farm businesses can be an invigorating part of the rural economy. I want him to see how it should be possible to feed people more locally, and to give him additional confidence, should he need it, that there is a strong demand for ethically produced food both here and overseas. I know that government is very keen on exporting, even though that seems rather at odds with the 'increasing self-sufficiency' line that many take; if we export more, we have to import to make up for it, given that we don't produce enough of the crops and animals that we can grow here to meet our own needs. So Tim talks about our sales overseas as well as here, and our delightful Italian pub manager, Valerio, asks whether he will be able to stay here after Brexit. We love having an international crew around us, and usually have interns and longer-term staff from all around the world. It feels good to have a more mixed community in rural England, where years ago we would rarely see anyone who wasn't born and bred here.

Then we set off around the farm and the first stop of course is the pigs. We walk across to see some newly farrowed sows and weaned piglets, stepping over the electric fences – I seem to have taken a rather convoluted route while absorbed in conversation – and he remarks on the differences between this and the indoor farm he had visited recently. He asks about food waste, and whether it should be fed to pigs. I give an unequivocal 'yes', as long as it is treated appropriately. Pigs have always been our waste recyclers, and it would be so much better to feed them on leftovers, rather than on the grains and protein crops that we could be eating.

As we get back to the gateway, I see that a sow and her piglets have escaped on to the road! So we round her up and direct her back into the field, an amusing moment as minister turns pig hand for a few moments. Before we leave he stops and looks back out over the paddocks in silence. I am not sure what he is thinking.

We drive the length of the farm to the agro-forestry, talking about definitions of productivity and the environmental benefits of organic farming. I really want him to get a sense of the potential of tree crops, and I hope that the work we have done at Lower Farm over the last 18 months will make an impact. In Barn Field, the almonds are in flower, providing a splash of colour on this grey day. He seems taken aback that we should be growing such a crop, so I show him the apricots too, and the tags on every tree that allow us to monitor everything about how they are performing. Then we walk down to Pump Field, where all this is on a much bigger scale. It's a forest of protective tree guards more than trees at the moment, though some of last year's plantings are peeping over the top of the tubes.

Then to the calf house, to see the calves with their foster-mums. But we are nearly out of time. I ask him why he hasn't been more vocal about the benefits of organic farming, and the response is that he feels it is not for everyone, and is concerned about lower yields. I agree on the first point, but make the case that the organic sector helps to lead the way towards much lower pesticide and drug use, reviving old knowledge and developing new techniques that will ultimately benefit all of farming and food. Just look at how many farmers want to learn about blackgrass control now that the chemicals are failing, or soil health; these days the Soil Association is inundated with requests for assistance from non-organic farmers who want to learn how we do it. On the second point, I suggest that more research would help to close the yield gap, but given that we agreed earlier that productivity is more about how we use scarce resources efficiently, and that approaches like agro-forestry can compensate for lower cereal yields, maybe he should be less concerned about this anyway.

He leaves, off to speak to the Swindon Conservatives. It's been a fascinating conversation, though he is largely inscrutable. I do hope that something has made an impression, and that he feels more enthused about organic farming as a result of his visit.

With the adrenaline draining away, I go to see the pigs. They are always good for me when I am in need of reflection.

The commune is empty, just straw blowing in the light breeze. Molly and her friends have finally been weaned. I climb over the fence into Flaxfield, and find the sows very happily chomping on the fresh grass in their new paddock. What bliss this must be after a few weeks in mud. Their udders are a little swollen, but that will reduce within a day or so as their milk dries up, and they come over to see me for their usual scratch and chat.

I look around for their piglets. They are at the top end of the field, near the badger's sett, mixed with some other piglets of roughly the same age who were weaned earlier in the week. The boar piglets are in one group, and the gilts next door, with a double electric fence dividing them to stop them reconvening. A bold blue stripe runs down their spines, David's marker to show that he has vaccinated them as they moved across in the trailer. They too are enjoying the fresh grass, and playing together, sometimes with a bit of a 'who's the boss?' scrap as they sort the pecking order with their new friends. Yet again, I am struck by the difference between pigs and cattle. Cows and calves hate being separated, however old the calf is; indeed it seems to get worse the longer they are together. But pigs seem entirely undisturbed by weaning, and appear to part from their mothers with barely a backward glance. A few of them wander over to greet me, but mostly they are in exploratory mood as they start the next stage of their lives.

13

Sex and the Pig City

24–30 March

Over the weekend, it starts to feel as though there's a glimmer of spring. As I walk up to the pig field, the clovers have finally got their heads up, and on the more fertile fields it almost looks as though there's something to eat. On an organic farm, nothing starts to grow until the soil temperatures are high enough for the leguminous plants to start moving, fixing nitrogen from the air to fuel the grasses around them as well as themselves. So while our neighbours are spreading manufactured nitrogen – 'sugar' as my father used to call it – we have to wait for the bacteria in the nodules on the roots of clover to do the job for us. This symbiotic relationship between nitrogen-fixing bacteria and plants that can host them is the main driver, along with recycled manures from livestock, of the organic farming system.

'Sugar' is a good name for artificial nitrogen. As the most

important nutrient for plant growth, nitrogen encourages lush development, but when it comes from a factory rather than the soil, it contains none of the micronutrients that are important for healthy growth. It's very like the empty calories we ingest when we eat sweets; they give energy, but also upset our metabolism if we overindulge. Too much nitrogen thins the plant's cell walls, allowing pests and diseases to take hold more easily, and fuelling weed growth too, leading farmers into a cycle of spraying chemicals, virtually all of which end up being proven – usually after decades of use – to be very bad for the environment and for us. It's an unfortunate merry-go-round. As farmers, we need to concentrate on feeding the soil microbes, just as we humans need to concentrate on feeding our gut bacteria. Bag nitrogen has allowed us to shortcut these principles that are at the heart of organic methods; our soils and our health have borne the brunt of these shortcuts.

So we organic farmers have to be patient. While surrounding fields are artificially green, we wait for warmth, for the bugs to start work. It's the time of year when it's possible to lose heart and yearn for an easy remedy. But then the magic starts to happen, and the nutritious clovers begin to grow, high in protein and omega-3 fatty acids, good for the animals, for us and the bees. In the autumn it's payback time for our patience in the spring; when all around, the sugar-fed grassland is depleted and lacking in goodness, our clovers power on, keeping animals satisfied and productive until late October or even November.

All is calm in the pig field. The sows seem blissful, appearing relieved to be shot of their demanding youngsters, and still munching their way through the grass and clover. They seem pleased to see me, and several of them come up for a scratch. There is one surprise, though. A lone piglet is hanging around their fence line; one of the babies who has perhaps decided that he doesn't want to grow up after all. I have a suspicion that he may end up back in with the sows if he can work out how to get in. I wander off to see Molly and her new friends. A few of them come over to greet me, but many are still feeding avidly, probably making up for their 18 hours on tight rations. Even those that approach me seem rather more tentative than was the case a day or two ago; I expect that the handling at weaning has made them more wary.

Weaning is a fairly straightforward operation, as long as the pigs are happy to be caught. David lures them with feed into hurdles attached to the trailer, which has been lowered to ground level. He then separates the boars from the gilts, putting a gate between them, vaccinating and marking them up as he goes. Then off they go to their new paddocks, boars one way, gilts the other. The sows are moved to pastures new as well. They load quite readily; they have been through this routine a few times already, and know that good things lie at the other end of the trip. The gilts follow on, learning the ropes.

On Sunday I catch up with David as he is underway with his early-morning rounds. As we chat, there's a commotion in the paddock adjacent to our weaned sows. A boar is becoming very aggressive with a sow; even from a distance we can see that she is lying down and screeching protest, yet he continues to try to ride her, bunting her hard to get her to stand. This sort of sexual aggression is unusual, but not unknown in pigs, as in the human population. It's hard to understand why it suddenly flares. It seems that early life experience impacts on the long-term behaviour patterns of pigs in much the same way as it seems to with people. Studies show that a barren environment and early weaning leads pigs to transfer their sucking and rooting instincts to the pigs around them, leading to tail biting and belly nosing, an extreme form of which is happening in front of us, triggered by sexual frustration. The boar could injure the sow badly, so David runs across to separate them.

*

Back at the house, our orphan pigs have turned into eating machines. Over the last week, they have doubled their feed intakes and are growing fast; the gilt, who we feared would never look like a normal pig, has transformed. Her back is now nearly straight, her tail tightly curled and she is almost as big as her brother. It just shows the power of good food, even after such a disadvantaged start in life; are there some lessons here for school food? They seem ready for some adventures!

On Saturday, Tim takes them for their first proper walk around the garden. Until now, they would follow him to the gate and then run back to their den, but today they venture out. Around the house they go, keeping close to their master, then a sprint across the lawn into the trees at the back of the garden. There's an old compost heap there, mostly lawn clippings, but they seem to think it's heaven, and start digging like little earth movers, black noses coming up for air every now and then. As they explore further, they come across daffodils, which I know are poisonous to many animals – but they seem to have an instinct about what's good for them, turning their snouts up and getting back to munching grass and digging turf.

Playtime in the garden becomes a thrice-daily event for Tim and the pigs. The fun and exploration seems to be stimulating their development; they learn to navigate the steep steps, and where the most exciting places are in the wooded garden; how to slip into the house when our attention is elsewhere, making a beeline for Dog's feed bowl and getting very cross when we stop them hoovering up his leftovers. Dog looks on in disdain. He is as good as gold with them, even though they are getting very pesky.

We are fascinated by the way they eat in the garden. They seem to filter feed, like whales, taking mouthfuls of earth, and somehow sorting it so that what they don't want falls out of the sides of their mouths. What they are actually eating remains a mystery; we assume beetles and earthworms and particularly tasty roots. They don't like to be separated, and will call in some distress if one gets carried away and ventures off from the other. And they follow Tim everywhere, heeding his calls. An easy way to lure them back to base camp is by tapping an egg, still their favourite food. Bananas are in close second place, though, and they are eating over a kilogram of pig pellets each day too.

They are incredibly sweet to watch now, bouncing around, full of energy, a far cry from the sad, malnourished creatures they were just a few weeks ago. It's taken a fair while for them to recover, especially hampered by the appalling weather, but it feels as though nothing could stop them now.

The start of the new week sees us turning the cows out for

the first time. It's 26 March, and we would usually expect them to have been out grazing a few weeks ago, even if just for a few hours a day, but as with everything else, the weather has prevented this; the ground has been so wet. During the winter, the cows seem content in their housing, but as soon as spring is in the air, they are restless. The clang of an opening gate, and they prick up their ears, bellowing in anticipation. When the moment finally arrives, they are off down the track like racehorses, and as soon as the earth is beneath their feet, their feet are in the air, kicking and bucking in joy. Such unbecoming behaviour from these matronly creatures!

I catch up with David and we go to see the poor sow who took such a beating from that aggressive boar yesterday. She is still looking very unhappy, and although she has had an anti-inflammatory painkiller, is reluctant to move from her arc. We give her water and feed where she is lying, and hope that she will start to mend.

Our boars are the only pigs that are brought on to the farm rather than reared here. They will have started their lives in an indoor unit, and unlike our pigs, their tails have been docked because most of them would have been destined for life on more intensive farms. Although we buy them as young as possible, so that they have plenty of time to adjust to their new free-range home, perhaps some of those early experiences are a factor in the kind of outburst we witnessed yesterday. Whatever the cause, if there's any sign of it from that boar again, he will have to go.

Molly and her friends have their curly tails back, traipsing around in the light frost and sunshine, perfectly at home with their new routines, it seems, albeit still a bit scatty. The boar group are already exhibiting boar-like behaviour, play-riding their pen mates, while the gilts are a little more decorous. The high-risk time of their lives is well behind them, and they will now concentrate on growing. At three months old they now weigh around 40 kg, and are gaining about 600 g every day. To sustain this growth, they must eat around 1.8 kg of feed a day, plus whatever they glean from the grass and soil.

Four days after weaning, and the sows are looking fabulous. Their milk has now dried up, and they are enjoying a last day of peace and quiet – not that they know this – as the boars will be joining them tomorrow when they should be fertile again. In normal circumstances, the sows should go into estrous about five days after they stop milking, when their energy levels rise as their milk dries up. I have been concerned that the very late weaning may have meant that the sows have had their first estrous while still with their piglets; by the last few weeks, the piglets were probably not drinking much from Mum, though it's almost impossible to know. So the sows may not synchronize as we expect. Two other sows have been introduced to the group, ones that failed to conceive first time around, so they have a second chance this week. The lone piglet who was hanging around their paddock has indeed managed to find his way in, and has made himself entirely at home. He doesn't seem to be suckling, or trying to, and he won't come to any harm there. The team have more pressing priorities than trying to capture him, anyway.

Our sow numbers have dropped a bit too far in recent months, so that now we have only around 180 sows in the herd. We aim to run at 200, which should produce over 3,000 finished pigs each year. Each sow will give birth, on average, eight times in the course of her life; after eight farrowings the numbers born in each litter tend to drop quite dramatically. This means we need to replace around 25% of our herd every year. The plan is to produce a few pure-bred boars that can then serve a select number of sows to breed the next generation of young female gilts. We therefore artificially inseminate (AI) a couple of the very best sows occasionally with semen from Saddleback boars who are unrelated to our sows. It's all designed to prevent any inbreeding, which, as is well known, can result in many kinds of physical and mental shortcomings.

Last year David's attempts at AI failed, possibly due to the lack of good semen storage facilities in hot weather. So we had no new young Saddleback boars, and couldn't use the old ones as they might have been serving their daughters. Incest is as bad an idea in the pig world as it is in the human one. Therefore we are short of young female gilts to replace the older sows who are leaving the herd. The Saddleback gilts in Molly's gang will be kept for breeding, but it will be another ten months or so until they have their first litter.

As an emergency measure, we have agreed to serve a few cross-bred gilts to keep the numbers up until the newly born Saddleback boar pigs are old enough to breed. In a week where we have no pigs to wean, and therefore no sows to serve, we introduce some of these white gilts to the boars to fill the gap. This week was one of these, and I'm pleased to see eight quite smart gilts in the boar pen. They make perfectly good mums, we've found, often having more piglets than the pure-bred Saddlebacks, but also being a bit inclined to lose one or two more too. It's that numbers thing again. Ten or 11 strong piglets are more likely to survive than 14 weaker ones.

David loads two boars, Larry and Leonard – yes, we do give the boars names as there are only 14 of them – from the paddock where they have been resting for the last few days. They are keen to move. They know the routine. These two have been reared together from piglets, so they will work together without squabbling. David backs the trailer over the electric fence, and the sows rush over to say hello. Out come the boars, and there is an immediate kerfuffle as introductions are done; lots of sniffing and snortling before everyone settles down for the evening. Well, everyone except Mrs Messy Bed. She is already ready for action, and Larry is happy to oblige.

Sows show that they are fertile and ready to mate by standing stock-still when the boar nudges her, or even when a human puts weight on her back. Courtship starts with the boar pursuing the sow, often frothing at the mouth, smelling her and pushing against her. The act itself takes several minutes; less than three minutes and the chances are that she will not conceive. Messy Bed and Larry seem to have achieved good harmony; let's hope she does a better job with her next litter than she did with this one.

Next morning, I am in the field early. The sows are all still abed, sleeping off their night of debauchery, and preparing for today's excitement. Messy Bed and Larry are lying close to each other, out for the count except that Larry has one eye open, watching but not reacting as I do my tour of inspection. It's my imagination of course, but I do believe he winks at me.

By the end of the day, three more of the group have been served, including Mrs Duroc. The boars have worked hard, and if they have another day as intense as this one, David will swap them with another pair to give them a break. Too many services and their sperm count will drop.

*

Having turned the cows out on Monday, it has rained ever since. On Thursday they churn up their paddock horribly, even though they are only outside for a few hours, and we decide that they will need to come back inside until conditions improve. This really is a terrible spring. Lambs are being born into cold rain, and becoming hypothermic very quickly. As chilled newborns are rescued from the fields the heat lamp in a pen we have set up in the calf house has an increasing number of lambs under it. But despite all our efforts, we lose more than we normally would. The pig field is a mud bath again; even the fresh paddocks are sodden.

The spring drilling is delayed too. We have some 100 acres of cereals to plant, plus grass seeds and turnips once the weather warms up. It's not a huge acreage to cover, but we haven't even started cultivations. If we didn't have so much that's exciting and challenging to keep us occupied, it would be easy to get morose.

We have been doing a five-year budget for our future farming plans, including the new dairy now that we have a more accurate idea of the building costs. Establishing this is a huge investment for us, and we need to be as sure as we can be that it will work financially. Having developed our business plan for this, we now need to work out whether taking on more land will help or hinder, and if we do go ahead, what arrangement would be best. We are all feeling a bit overwhelmed by the need to make so many decisions in such a short space of time, and by the amount of work this is going to take. At the same time, it's putting our embryonic family partnership into action, and that feels good.

One decision we make is that we will take on the management of the new land for the summer. The farm manager there is leaving this week, and there is no one to take on this role. With calving and lambing underway there, and crops to drill, we have been asked if we could oversee things for a time, while longer-term plans are developed.

With my head buzzing, I walk the Downs in fine drizzle, the ground soaking underfoot, even up high on the ridge. Three red kites circle around me, and three hares take off across the downland pasture. I drop down to the pig field, and slop across wet tracks to see the pigs. It's raining heavily now, but Molly and her fellow weaners are all out and about. As I stand amongst them, there's a sudden commotion as two very funny-looking ducks fly over and land close by. These two have been hanging about here for a couple of years, and I'm rather embarrassed that I don't know what species they are. They look like ruddy ducks, but I thought that they never leave water, and are, in any case, very rare. Whatever they are, they provide the excuse for some mock hysteria, with pigs running around as if it were a drone strike approaching.

In their paddock a few rows away, the sows and boars are sleeping, however, all sensibly tucked up out of the rain except for the lone piglet who is rummaging around the arcs, then pushing his way in amongst the adults to find a cosy spot for a snooze.

One of the boars stirs and grunts in annoyance as the impudent youngster invades his personal space, and I'm rather taken aback by the piglet's response. Instant submission would be the obvious course of action, but he stands his ground and responds with a challenging grunt of his own. The boar is too sleepy to take things any further, but this piglet is pushing his luck, it seems to me.

It is Easter weekend, and David is leaving for two weeks' holiday from Good Friday evening. He spends a chunk of the day catching up with overdue record-keeping, and then has a final handover session with Tony and Rachel who will be keeping the show on the road while he is away. We all meet up in the field and discuss the work plans, given that there is still plenty that is behind schedule. While we chat, another of the sows is served; it's 314, one of the gilts. David confirms that there has been plenty of activity, and that he thinks that all except one of the sows have mated. So my concerns that they may have come into estrous before weaning seem to be unfounded. With any luck, this group will farrow together again in late July, an easier time of year for them and their piglets. Before I leave, I pay them a visit. Most of them are outside, rooting away, while one sow and a boar are stretched out dozing in the straw, noses almost touching. Of course, it's Messy Bed, the lazy hussy!

14

How Much Land Does a Man Need?

1–13 April

There's a glimmer of sun on Easter Day; is this an April Fool? I walk the Downs to see the young heifers who were turned out a couple of weeks ago to make space in the barns for the cows that were about to calve. They haven't had the easiest time of it, with snow and wet weather, but they look bonny enough. They went to the bull a few months ago, so we will be pregnancy testing them soon, in the hope that most of them will have their first calves in the autumn.

From the hill, I look down on Sophie doing her morning lambing check, buzzing about the fields in the open-sided Polaris. At one point, she leaps from the little vehicle to catch a ewe; I can't see clearly enough from here, but it must need help. She is down on the ground with it, and five minutes later has delivered a lamb. As Sophie moves carefully away, the ewe starts to lick her newborn dry.

When I get to the sows, I'm intrigued to see one of the boars pacing the fence line, frothing at the mouth in the manner that means he's tracking a fertile sow. But the sows next door are all due to farrow, so surely he's got the wrong end of the stick.

Molly's group are more friendly today, and it's good to renew our acquaintance properly. I'm especially pleased to see that the little Saddleback who has been the most underdeveloped in the gang for the last few weeks is looking stronger. In the end, we didn't put any of the poorer piglets into 'special measures'; they have stayed with the sows so long that they are at least as big as most normal weaners now, and anyway, the pig team have had so much on.

By the next morning it is raining again; the sunshine *was* an April Fool. I meet up with Rachel and Tony, who have a busy week ahead. A couple of sows have farrowed unexpectedly while they were still in with the dry sows in their big communal arcs. We try to avoid this, especially if there are quite a few sows together, as there is a higher risk of piglets getting squashed by the cumbersome, heavily pregnant sows. So far, these litters are fine, and they are in arcs with plenty of room, but will need moving soon. There are new paddocks to set up, farrowing paddocks to divide, dry sow troughs to move, and litters to wean. And with the renewed rain, more bedding needs to go out too, and that, of course, is on top of the usual feeding and checking.

Henry and Dai have been taking the reins at the new farm over the holiday weekend, and they have made great progress, meeting with the staff, doing night checks on calving cows, mending bits of equipment that have been languishing for a while and even getting all the records on to the computer. I'm very impressed, especially as there is plenty going on here as well.

We meet up in the pub in the evening to say goodbye to a wonderful Brazilian intern who has been with us for six months, and over beers and Mojitos (her favourite), Dai reflects on what a good team we are together, with all our very different skills and attributes. It does feel that as a gang we can rise to the challenges and opportunities.

And there will be plenty of these ahead. The biggest challenge for us and other farmers is not having much idea what is around the corner. Weather is always our greatest unknown, and that is likely to become even more unpredictable with climate change. We have no sense of what sort of trading arrangements will be in place post-Brexit, nor whether we will be able to employ people from overseas. We don't know what 'public money for public goods' will mean for us and how the changes will impact on land values and rents.

Standing back from our own situation, I spend much time thinking about what I would do if I were able to wave a magic wand. Humanity's impact on the natural world, on which it ultimately depends – though we often seem to forget this – has been immense, and much of this impact has been through the way we produce our food. Even before we became farmers, we drove many species to extinction, but as our numbers grow, we are ever more destructive.

Despite current and projected numbers of people on the

planet, we could do so much better at reducing our impact. We *can* sort the energy problem; the technology is there, or nearly there, as renewables and batteries get ever better and cheaper too. We *can* recycle or re-use almost everything, if we build it right in the first place. These things are just about political will, standing up to vested interests and making the right choices about where we put our efforts and money. I know that's a big 'just', but it is possible!

Land is where it gets complicated, because today that seems to be the limiting factor. It may seem as though there's plenty of it, with huge tracts of land across the globe where population is very low, but that's largely because of a lack of good soil and water. So the most sustaining environments have been overworked, their soils eroded as they have been cleared of trees and shrubs, prairie grasslands ploughed to make way for crops, and all the other species that have depended on these places died away. Alongside the huge amounts of carbon released into the atmosphere, and the reduction in the ability of soil and vegetation to soak it up again, we have got to the point where the approximate weight of humans on the planet is 350 million tonnes, our farmed animals 1,010 million tonnes, and non-marine wildlife only 40 million tonnes. We and our livestock have squeezed out most other species.

In little old Britain, the tensions play out very clearly. Everyone wants land, to build new homes and roads on, to make a living from, to hide away on, to play on, to protect nature on. We farmers seem to have bought the idea that our moral duty is to feed people, even while the market value of what we produce is so low that we find it hard to make a living without public subsidy. It seems at times like an excuse to keep doing what we know and love to do, and what many farmers like best of all is to drive big machines, and avoid too much complexity. Egged on by the many companies who want to sell us products, we have ended up with a denuded landscape of large fields and over-grazed hills, producing a few commodity crops, much of which is fed to animals that are confined indoors in ever greater numbers.

Organic farming is a significant step in the right direction. The avoidance of artificial fertilizers and chemicals means that we have to follow time-honoured ways of building fertility, through legumes and rotations of crops and animals. Forbidding the routine use of antibiotics and other drugs makes us focus on good husbandry, and, of course, our animals must be free-ranging whenever the weather permits. This all brings benefits to soils, water, wildlife and the nutritional content of food. But it's not enough. To meet the challenges the world is facing, we need to do much, much more.

Natural systems are very complex, with infinite interdependencies, and usually very productive too. Our current ways of farming, even organic ones, aren't perhaps the most efficient ways of generating biomass and foodstuffs. We need to shift from simplicity to complexity, mimicking and working with nature to produce what we need. In my view, there's no problem with our ability to produce enough food, and other products too, if we move to more ecologically intensive methods. As shown by my own experimenting at Lower Farm, I've long been interested in the idea of permaculture, where landscapes are designed to deliver a range of benefits, and the food-producing components are more focused on perennial crops, like shrubs and trees, rather than annuals such as cereals. It's been hard to find large-scale examples, though, and it's some of these that we need in order to convince policymakers and farmers that these methods may be a much bigger part of the way forward. Hence our attempts at Eastbrook to begin to explore elements of this approach.

The problem currently is the economics of these systems as the diversity increases harvesting costs, and makes processing and marketing less straightforward too. These techniques need people, and could provide much meaningful and enjoyable work for the many folk whose livelihoods may soon be displaced by driverless vehicles, artificial intelligence and robots – albeit that the thorny problems of incomes and housing need to be tackled too. These systems also need processing and distribution infrastructure, ideally locally so that healthy foods can reach consumers with minimal packaging – another of our great challenges.

At the other end of the scale, in some people's eyes at least, is the rapid development of novel crops. Algae, insects, seaweed, fungi and laboratory-produced meat, have little reliance on land, and may pass the test of energy, nutrient and water efficiency. This is also true of more novel farming techniques such as 'vertical farming', where crops are grown in stacked layers, under LED lighting, and hydroponics, by which plants are grown in water. If more of our food is produced this way, will the tensions over land become less? I think so. Before long, a mix of ecologically intensive, perennially based cropping on the one hand, and high-tech approaches that are very resource efficient on the other, could allow much more of our land to be free to provide eco-system services like carbon sequestration, water management and plenty of space for other species to thrive, for people to enjoy, and the ultimate heresy, for people to live in.

Like many rural dwellers, I've long been nervous about more development in the countryside. I love the peace, and hate seeing green fields turned into housing estates. But I now recognize that well-designed accommodation, with space for people to grow some of their own food, and with nature plumbed in through trees and wildflower meadows alongside water recycling and renewable power, could have huge benefits for the many people who would like to live this way, and for nature too. A rather sterile arable field can support more wildlife and potentially grow more food when appropriately developed for housing and smallholdings or allotments than it ever does as a monoculture crop. More people in the countryside could be part of the solution, infinitely possible in a world where home working is easier for many (broadband permitting) and with opportunities to get involved, full- or part-time, in the food revolution we need.

Where do farmed animals fit into this picture? Well, my pigs would be thriving on the offcuts and waste from local food-processing, with much less dependence on grains and protein crops that we could be eating directly, rather than feeding animals with them. I wouldn't be keeping as many of them in one place either, and their meat would be sold much more locally on the whole. We would still be producing milk, and some beef and lamb too, from grass up on the Downs, or on the heavy clays that are unsuitable for cropping – though we might stop thinking that this land is meant to be an open plain, and allow some regeneration or plant more trees there too, as we are starting to do at Lower Farm. There would be many opportunities for people to keep small flocks of poultry, a good way to supplement incomes on modest amounts of ground, or to keep bees, a rather undervalued livestock, or start small-scale food businesses. Some of these would grow, some might choose to stay at a more artisanal scale.

Diversity is the watchword, in business as well as in ecology. There would still be broad-acre farming, producing cereals and potatoes, but at the very least it should follow organic principles, protecting soils, rotating cropping and avoiding pesticides. In the uplands, which are less important for food production but vital for water, carbon and wildlife, farmers would be paid to manage the land to secure these benefits, with modest numbers of livestock that support rather than destroy a varied and less denuded countryside, one that is attractive to visitors and therefore with increasing hospitality opportunities for rural communities. In the west, where the high rainfall allows us to grow grass very efficiently without irrigation, grazing-livestock are the best option for fertility building, while in the east, the drier soils would be perfect for pigs and poultry. The land there is crying out for the organic matter and the fertility that free-ranging herds and flocks would provide, helping to sustain high yields of the arable crops that we need long into the future.

*

My pigs might well like to be transported to a drier land, or at least their keepers might. The endless routine of bedding-up gets wearing, and we are using so much straw that we are fearful of running out. We can't make any more until harvest, and we need to have enough to keep the pigs comfortable till then. It is so wet underfoot that we abandon our time-honoured tradition of hiding chocolate eggs in the stunning Eastbrook Valley, and instead use the Ridgeway where the sealed surface makes it easier for families to get about.

Our orphans are growing daily, full of confidence and just so mischievous. They have the run of the yard during the day, and they make sure that their presence is never in doubt by establishing their latrine right by the back door. Getting in and out of the house becomes a game of minefield, both dodging the poo and trying to stop the troublesome two from barging their way into the kitchen. When they do make it in, it's a race to the dog's bowl; if we win and sweep it out of range, then they become vacuum cleaners, snuffling up any debris on the floor. There's always plenty of that in our kitchen!

Their favourite foods are still eggs and bananas, though, and we use their passion for these ingredients to good effect. Tim, and guests too, take them for ever longer walks, and as their confidence grows, it feels as though there's an increasing risk that they may just hot foot it off around the village, so we need to keep them responsive to our calls by offering food as a reward. They used to be nervous of new surfaces, be they puddles, or snow, or tarmac, or decking, but now they have met all these, so once the yard gate is open, they are away. Up the slippery decking steps, under the tables, down on to the lawn and a scamper to first base, the old compost heap. Next stop is the trees; they would happily spend all day nosing into the soft undergrowth, peeling back the surface like seasoned professionals. Then one will decide that it's time to check out the mozzarella factory, accessed from the rear of the garden. He or she will sprint off, and then the other will chase along after – such a comical sight with their bouncing enthusiasm.

At this point it's time to reel them in; I don't think that the Environmental Health Officer would be too thrilled by having pigs in there. They so far respond well to our voices, whether it's my 'Come on, little piggies' or Tim's 'Chop, chop', and scootle back to us with touching and relief-inducing reliability. Tim has been confident – or perhaps foolish – enough to take them down the road to meet our neighbours. They seem a little unsure of other pedestrians, but scuttle past and stick close to their master's trousers. I think this is the fulfilment of a long-held dream, a couple of devoted pigs to follow him about.

Like any good parent, he is starting to think about their future. 'I think I will train them to be truffle hunters,' he muses one morning. 'I can seed your new trees with truffles, and then hotel guests can rent the pigs to search them out.' Another innovative idea from the ever fertile imagination of Mr T. Finney. I suggest the pigs might be getting on in years by the time we have any truffles for them to find, and perhaps – if he really doesn't want to eat the pigs – a more realistic option would be for the gilt to be kept for breeding, and the boar to be vasectomized for use as a teaser for the sows. They both seem perfectly attractive options for the pigs, especially when only a few weeks ago they didn't look likely to have any future at all.

Indeed, I would be happy to breed from the gilt, despite her unpromising start. Her ears are still too big for her body, but the ratios are coming into line, and from a hunched runt, she has metamorphosed into a sleek, muscular creature, with the length of body we look for in a breeding animal. We know that her dam was very fertile, and a good mother as well, doing her absolute best for her offspring even when she was struggling herself. These are characteristics that are likely to be transmitted to her young.

As for the boar, he doesn't need to look good for the future I have in mind for him. Not that he's ugly – far from it now. Every year we vasectomize a few male pigs and keep them close to or in with the gilts that we want to breed from. The presence of a boar will encourage them to cycle, and allows us to see when they are coming into estrous. If we want to artificially inseminate some of our best sows too, with pure-bred Saddleback semen from an unrelated boar, then the V boars, as we call them, will help us catch them at their most fertile moment.

For now, the more pressing challenge is finding them somewhere more appropriate to live. With spring weather around the corner, we hope, and the pigs growing at such a rate, they need to have their own paddock. We start planning to bring an arc into the garden woodland, but then Tim has a better idea. We have a bit of rough ground near to the pub and farm offices where we reared turkeys for a year or two, but is now rather overgrown. This would be the perfect place to rehouse them; they could dig the patch over, so that we can reseed it in the autumn or maybe even start to grow some vegetables there. A good plan, but for now it is still too wet, so we continue to be greeted by squealing when we arrive home – they seem to recognize our cars as other visitors don't get the same welcome – and by pigs jumping up on to their hind legs against the gate, pooing on the doorstep, and pleading for bananas. Perhaps they could hunt bananas rather than truffles.

15

Rain, Rain, Go Away

14–18 April

It stops raining for a day or two, and our spirits lift. Finally, I can walk the field without wellingtons, and the grass is growing back in the paddocks that the pigs have vacated. Henry joins me amongst the pigs on a hunt for water leaks; we have run out of water at Starveall, so there must be some escaping somewhere, and it should be easier to spot now that there's not water everywhere! The pigs use a lot of water, especially in the summer when we have to create wallows to keep them cool, and it's very easy for a water fitting to come loose, or a trough to get upturned, then suddenly the reservoir is struggling to keep up with demand. There's been enough water falling from the sky to make a mess of the field, without a leak adding to the mud problem.

I make the most of better conditions underfoot, walking some fence lines and picking up scraps of the black plastic silage sheet that have blown from the clamp over the winter. We have big bags in the pig field to put string and netting from the straw bales in, and I stuff my collection into one of these before heading off to see Molly and co., and to catch up with Tony and Rachel. They are making the most of a drier day or two to re-bed everything, and given that the forecast is still not good, to create straw tracks to the feed and water. Although the feeders are all on railway-sleeper platforms, the routes the pigs use to get to them have become boggy, and putting some bedding on these should encourage them to keep eating.

Some of the big arcs that house the growing pigs are marooned in mud, so they move the worst of them, resetting them on more solid ground. There are also some cull sows to go – ones who have failed to get in pig again, and so are off to the land of sausages – these need to be caught and put into the Forty buildings ready for loading. There's plenty to do, and even though David's away, having Rachel on board is allowing us to keep apace with the workload. I'm very encouraged too, by the way she is keeping up with record keeping. This task is a constant source of irritation, and I am always nagging the pig team for the data; we need to know exactly what's going on – how many are born, weaned, any deaths or veterinary treatments – and so often we are weeks behind getting these details on to the computer. If she helps us solve this problem, she will have a job for life!

Molly and her old and new friends are loving all the fresh straw that's piled up outside their arc. They use it as a mountain, jumping up on to the highest peak; they play hide and seek around it and then, of course, they dig in it, wriggle into it and lounge on it. We find the same with children at the pub; a few bales and some loose straw and kids seem to be happy for hours. As a child, I myself would make bale houses in the harvested fields with my sisters and friends. The big bale hadn't been invented then, and the small ones were light enough to lift out of the formations of eight that the baler sled left them in to create these homespun playhouses. When all the bales were safely stacked into barns for the winter, these too became our play zones, climbing up the stacks in a way that parents and authorities would have a fit about today. Straw is just such fun!

We have a big dairy PD (pregnancy detection) session with the vet today. This is the moment of reckoning; has the winter feeding regime been consistently good, so that the energy status of the cows was sufficient to maintain their fertility? Have Teo, our herd manager, and his assistant, Andy, watched the cows closely enough to spot when they are in estrous? It is vital that a cow has a calf each year, and we want them to calve within a fairly narrow window of time so that we can concentrate all efforts on getting every one safely delivered. It's much more worth your while to get up at 2 a.m. to check them if there are several likely to calve, and no one wants to get up at 2 a.m. for too many months. So we have tried to get them all pregnant over a nine-week period, which is tough to do without using hormones to help synchronize estrous – and, of course, we cannot do that as organic farmers.

It's a great result. Out of 138 cows, 120 are in calf, and from 44 heifers, 40 have held. Many of the foster cows are pregnant too; as these are older cows, we expect a lower conception rate.

Teo and his team have done brilliantly; we will finish calving by the end of November this year, three weeks earlier than last season. That means that most of them should be able to calve outside in September and October – they always calve more easily in the field, where they can find their own quiet spot – and we will be up to peak milk production quickly too, at the time of year when the price of milk is highest. An added benefit is that Teo and Andy will have a relatively peaceful Christmas, with no broken nights calving cows.

We have just heard that the first load of the Irish heifers that Sophie and Dai have bought are arriving tonight. We need to quarantine them, and the best place to do this is at Lower Farm, a long way from any of the rest of our cattle. But it is still very wet down there, and with all the tree planting, there's much fencing still to do. We have selected a field that is not too sodden, with as much grass as there is anywhere, and hope they don't churn it up too much. They are due to arrive between 1 and 3 a.m., so there's another disturbed night ahead for Sophie and Dai.

*

I head down to Lower Farm this Saturday morning to see the new arrivals. They are restless after their late-night adventure, and some of them look as though they could do with a good feed. I leave them to settle down and walk across the adjacent field where Ben, who is Head of Horticulture at the Soil Association and oversees our agro-forestry project in his spare time, has been planting some specimen trees to create a more traditional parkland feel in front of the house. These will all need to be individually fenced, so that we can graze the field without the stock damaging them. The stream runs on the east side of this field, and I walk its course down to the Botswicky woodland. The willows have now been pollarded, and some of the brush has fallen into the stream, slowing the flow and creating some cul-de-sacs in the oxbows. This is already leading to rushes and other vegetation beginnining to grow in the quieter waters, and I can imagine that soon more wildlife will start to benefit from these mini-wetlands. Again, I feel that I am learning all the time. The received wisdom has always been to get streams and rivers flowing as fast as possible, to get water off the land and away; but now we realize that if we are to manage the risk of flooding downstream, and provide more habitats for wildlife, then we need to slow the flow, not accelerate it. This may not benefit us as farmers, who want our fields to dry as fast as possible, but the wider advantages are indisputable.

This challenge, of how to help farmers like me to learn more about ecology and how to manage it, to help wildlife thrive, and look after the water that runs through our land, is one that needs a lot more attention if we are to fulfil our role as custodians of the countryside. We are not taught this to any degree in our agricultural courses, and while a few farmers may have an especial interest and seek out the knowledge they need, too often the cry goes out 'we want to be farmers, not park keepers', which I think stems mostly from feeling out of our depth. It takes a deep understanding of natural systems, of the life cycles of many species, to manage land for the multiple benefits it needs to provide, knowledge that it is often assumed farmers have – but why should we know this? Even when farmers have been encouraged into environmental schemes, to plant wildlife habitats for instance, the prescriptions are laid out to such an extent that it can feel like painting by numbers; we are told what to do by the experts, and if we don't do exactly as prescribed, chunks of money are withheld from us. So we don't have any incentive to learn, and it's possibly safest financially not to know too much, but just to follow the rule book.

Today I'm frustrated by one example of this in the work we are doing to transform Lower Farm into both a far more biodiverse and more productive patch of the countryside. The willows that we have pollarded should apparently have been coppiced in order to receive the grant to undertake this expensive work. What's the difference, you ask? Well, coppicing takes the tree right back to ground level, so that it regenerates from there; this is best suited to younger trees, before they have developed a strong, mature trunk. Pollarding involves cutting the branches back to the main trunk, so that they shoot again higher up. Our willows are mostly old, and have been pollarded previously; cutting them through their huge trunks seems an act of vandalism that might even kill them (though willow is notoriously hard to destroy), and would change the landscape dramatically for many years. More importantly perhaps, given that the objective is to stop the trees splitting, and to engineer a dappled shade over the stream that will benefit its inhabitants, taking the trees out completely seems entirely misguided. Our tree advisers agreed that pollarding was the right thing to do, so that's what we have done. But in doing what seems to be right for this circumstance makes us ineligible for the grant that would have paid for a proportion – though by no means all – of the cost.

A similar situation has arisen concerning our protection of the newly planted woodland. We applied for normal cattle fencing and big, high, expensive deer guards for the trees, then realized that it would be more attractive, more effective and cheaper overall to deer fence the little woodlands, so that the trees could be spared the five-foot tubes. We would be claiming less grant, and doing a better job. But it looks as though our claim may be rejected, again for trying to do the right thing.

Moving to a more 'outcomes'-based approach makes much sense in many ways, so that farmers are paid for their results, rather than for following the rule book. Attractive though this notion is to me, given that it would stimulate us to learn and experiment, focusing on what we are trying to achieve – for example more skylarks, better water quality, or higher soil organic matter – it's not without many challenges too. Skylarks move, and may be hard to count; water runs across many farms, and soil organic matter builds over many years, and we don't have a lot of data on how much there is to start with. So we may need a mix of approaches, combined with thorough monitoring so that we can build a clearer picture of which methods are working best to achieve the objectives we seek. At the same time, we need to help farmers learn new skills, and to be as proud of their role in enhancing nature – as some are already – as they are of their skills in growing a 'clean' high-yielding crop of wheat, or persuading a cow to give ever more milk.

It's the same debate with animal welfare. At the Soil Association, we know that it's not enough to just have the highest standards, the prescriptions of what you can and can't do – but also vital to ensure that the management and husbandry is helping the animals have a good life. The skills of the stockperson make all the difference in enabling the potential of the system to be realized in practice. So Soil Association inspectors will check on key welfare indicators, like lameness levels, skin condition and feather loss – this one doesn't apply to pigs! – to ensure that the husbandry is as excellent as the circumstances in which the animals are kept. If a cow is lame, being free in a field and having to walk back and forth to the milking parlour may be worse than being shut inside. One of the things that I am most proud of is that this way of assessing welfare outcomes is now being used on many farms, not just organic ones. And, of course, knowing that Soil Association farms not only have the highest standards for the freedom and space that our animals can enjoy, but that we are also checking that they are in a fit state to enjoy it too.

Given all these complexities, it's maybe not surprising that Defra officials and their political masters are having some difficulty in developing the best way of supporting farmers to deliver more than just food in our soon-to-arrive 'brave new world' outside of the Common Agricultural Policy. And it's why I feel so strongly that backing systems like organic farming and agro-forestry, which have so many inherent, scientifically proven benefits, should be encouraged and supported in the future. There's still much work to do to keep improving our approaches, and investment is required there too, but if the goal is to produce sufficient quantities of healthy food, while caring for our farm animals, wildlife and soils, it seems crazy not to fully get behind these agro-ecological systems.

*

This afternoon I drive up to Starveall, and walk with Dog towards the Downs. I'm less worried about disturbing the ewes now; many of them have lambed, and are well mothered-up. They do look bonny! Skippy lambs are almost as joyful a sight as naughty piglets. Across the track from the ewes, Henry is on the plough, turning over the stubbles on Hedge End. At last, we are moving closer to planting. The stubbles have been a great resource for birds over the winter; organic stubbles are especially beneficial, as there is more weed in the crop and our ancient combine harvester possibly leaves a bit more grain on the ground than it ought to. Seagulls follow the plough looking for the worms it will reveal, and the brown, loamy soil smells delicious.

I stop to see the pigs on my way back. The sows and boars are very quiet, with only one of the females outside. She is the gilt, 312, the one who could not bear to be parted from her friend, 314, and I can see from the blood and mucus around her rear that she has just been served. All of the piglets are out and about, however, and they are on very jolly form. The drying ground seems to make rooting even more fun, and when I sit on the straw heap outside their arcs, they all come over for some fun and games. With my boots dangling down, they set about cleaning the soles, and trying to join me on my comfy seat. I'm in no rush to leave.

It's a short-lived respite, however. By Monday it's raining heavily again, and by Tuesday, when I spend the day showing a group of Defra officials around the farm, the pig field is a quagmire once more. Many of the group are new to the department, which has drafted in hundreds of people to assist with the numerous tasks involved with leaving the EU. Few of them have spent any time on farms and so the Soil Association and I have hosted a number of visits for groups to help them become more familiar with the issues first hand.

My guests are here to learn about farming, especially organic methods. I debate with them the relative merits and problems of keeping pigs this way from an environmental point of view. On the plus side, virtually no antibiotics are used and the birdlife has had a great time here all winter when the surrounding countryside has been bereft of food; on the minus side, we have churned up the soil with our tractors, creating horrible ruts that will take a while to heal; there is a risk of nitrous oxide release from saturated soils with all that manure from the pigs; and we have used more feed than we would do if the pigs were inside, not running around, having fun and using energy to stay warm. Of course, my highest priority is the welfare of the pigs, and I am still convinced that they would choose to be here rather than on concrete, despite the muddy conditions.

Nevertheless, rather depressed by this endless rain, I yet again consider whether we should experiment with a more indoor system for the growing pigs during the winter. Could we create bale pens, or tents with a woodchip run outside, and a concrete sleeper-track along the front to allow us to feed and bed without damaging the ground? Perhaps we could feed them more silage or waste veg to give them the benefit of some green material as well as their pelleted feed while there's little grass growing. And it would be easier to make use of the whey in a more contained site. The main problem is ensuring that we don't leak nitrate from this more intensive approach, and that the pigs stay fully occupied and happy, with plenty of rooting material. It's hard, however, to think where on the farm we could do this. Starveall is the obvious site, but it will hopefully be home to a new herd of cows by next winter.

Over lunch in the pub with the Defra team we discuss meat-eating, and whether attitudes are changing. This group are mostly in their 20s and 30s, and it seems that they and their friends are aware that overindulging on meat and dairy is environmentally irresponsible. This is a well-informed and educated bunch, though, and possibly not representative of the whole population; it may be a while before the majority of people think that less but better-quality meat is the way to go.

*

My depression deepens on Wednesday. Is this the worst day of the winter, with low cloud and fog, and more rain after so many inches during the last three days? The worst, not because it's the coldest, or the greyest, or the wettest, but because it's just dragged on and it should have stopped long before now. By Friday 13 April it's still horrible, but despite the superstition of the day and the ongoing gloom, my mood is lifted by meeting up with Tony and Rachel in the field. They have cracked on with the work this week, weaning two pens, sorting out all the fences so that the escapees are back where they should be. They have moved all likely-to-farrow sows into proper accommodation and have given big arcs to sows with older piglets, so that they can establish cosy beds without the constant bedding-up that the small farrowing arcs require. And when I get to the office, I find all the records up to date. Impressive, under the circumstances.

Henry, Sophie, Dai and I meet up late in the day to review where we are on all our ventures. We are waiting to see whether our landlords are prepared to help with the investment at Starveall, especially getting electricity to the site. I have been speaking with their agents, and they have promised an answer next week. What should we do if they say no? Can we afford to spend all this money ourselves? It's a huge investment, and while our business plan says it makes sense, we are in a time of such uncertainty with Brexit looming, and with bovine TB too, that there are a number of things that could upset our careful plans.

Henry and I are nervous about reducing the size of the pig herd until the new dairy is up and running, and starting to pay back its debts. We are keen to keep our eggs in at least two baskets, one of the approaches that has served the farm well over the decades. We debate this, at times a bit heatedly, and come to a compromise; we will maintain the herd at its current size until autumn 2019, and re-evaluate then. During this time we will try to find ways of overcoming some of the downsides of keeping pigs the way we do, such as winter housing or finding some lighter land to keep them on during the wetter months, in order that we can reduce the area and impact on the land around Starveall. I'm also keen to get going with developing a dairy brand, so that we are not so vulnerable to any future swings in milk prices. The organic milk price has been stable for many years, but there are no guarantees that it will stay that way. I'm relieved by the decision to maintain pig numbers for now, and Henry and I feel much more confident about making the dairy investment knowing that we have the pig income both on the farm and from our marketing business to help us through this period of change.

On Saturday, at long last, the forecast is for fine weather. I walk the Ridgeway to Idstone with Dog; he is slowing down noticeably now, and doesn't enjoy the long walks I want to go on. He is 13, quite old for a greyhound, and while he has mostly recovered from his funny turn a few weeks ago, I'm very aware that he won't be with us much longer. After a gentle mile or two with him, I pop him back in the truck and set off across the pig field.

The sun is trying to break through an early-morning mist, and as I approach the pigs, the only sounds are bird chatter, the odd clang of feed-hopper lids, and the squelch of my boots in the saturated ground. The gilts come across to see me, squelching away too, and seem full of beans. Indeed, I'm rather taken aback by how much girl-on-girl action there is going on. The boars ride each other from a young age, but I've not noticed gilts get quite so frisky with each other before. They are growing very fast now, though it should still be another two months before they come into estrous for the first time. The boars are less friendly, seemingly more intent on feeding and rooting and playing in their straw heap than socializing with us humans. There is a clear difference between the sexes now.

The sows, however, are very engaging this morning. Almost all of them gather round me, and I can see that they have gained some condition – not that they were thin when weaned – and they certainly don't need to get any fatter. They are very vocal, avidly chattering away to each other in grunts and low squeaks that I wish I understood. It does sound like conversation. This is not dominance-related behaviour, or the warning barks of the nervous mother, or the mock hysteria of the startled group, but a gentle chat between old friends. Leonard and Larry are with them, and Leonard lies down in the mud, scooping into the most liquid patch with his long nose, and then rather comically wafting it about like a giant ice-cream cone. Mrs Duroc, sleek and chubby, sidles up to him, and tries to reach some of this especially favoured slop.

I can't resist going to see how some of the newly farrowed sows are doing. There are lots of piglets out and about on this mild morning, and I find that a young white gilt who became pregnant by accident rather sooner than she was supposed to has given birth over the last few days. Her piglets are out of the arc, and I think I can count seven, which is a perfect number given that she is a teenage mum. She should be able to feed them well and keep growing herself too. She trots energetically along the strip of her paddock towards me, and then stops, ears pricked. She snorts and wheels around, running back to her babies, bucking and weaving like a piglet herself. She noses them back into the arc, away from scary predators like me, and then comes back for another look at the scary thing. Once her back is turned, out come the piglets from their arc again, then back races the gilt, barking her warning. This could go on all day, so I leave her in peace.

The day heats up, and by afternoon it is over 22°C. You can almost hear the grass growing, and even the lambing fields, which were starting to look too barren to sustain the ewes' milk flow, suddenly look green and luscious. Tim and I play with the orphans in the garden, and all the heartache and worry seems worthwhile as we watch them running about, confident and healthy. They decide that a dip in the middle of the lawn, where an old tree died, is worth a serious excavation. A few minutes later we have an opencast mine, before they move on to further destructive duties. Dog gets fed up with their antics, especially when they involve barging past him, sniffing his tail or trying to invade his kitchen, and having been the epitome of refined disdain until now, he has a good snap at the upstarts. A bit more respect is required for Dog, please! I think they have got the message, but I'm also not sure that they will remember it for long.

On Sunday, it's grey again, with some rain in the air, but the warmth of yesterday, and the promise of more next week, keeps spirits up. I walk the lunar landscape of the now largely empty Broad Gap, across the drying mud, to see how the older growing pigs are doing. There is a distinctly deep pinkness about them after their day in the sunshine. Just like us, they have made the most of the fine weather. They look good overall, though there are one or two cauliflower ears and the odd bit of head shaking, which suggests that there's a hint of mange around. We need to get on with treating the younger groups; the sows are already having a treatment as they go into their farrowing paddocks. These groups are too close to slaughter to treat, as there is a withdrawal period that must be observed, a much longer one for organic pigs, to make doubly sure that there is no possibility of residues in their meat.

I also find a hopper that is spilling feed very wastefully, and another one with no lid. It seems we are not entirely on top of everything yet.

*

On Tuesday, the Soil Association hosts an open meeting for farmers at the Royal Agricultural University near Cirencester. The main purpose of this is to engage producers in the government's 'Health and Harmony' consultation – the document looking at the future support arrangements for farmers once we leave the EU – and to make sure that our response reflects our members' views. But it is also a great chance to find out what is on farmers' minds – apart from the dreadfully late spring, shortages of forage and straw, and worries as to whether any crops will get planted this year. I raise my concerns about out-wintering livestock, both cattle and, especially, pigs, in these seemingly ever-wetter winters, and find that a number of other farmers are similarly worried. We all have the same conundrum. Animals stay healthier outside, on the whole, but they do make a mess. It's certainly time for some innovation, an approach that will give the pigs what they most want – soil, straw and space – while stopping the damage our vehicles are doing in the course of servicing their needs. It's a big job to develop this, and the farm team are too busy to give it much attention. I wonder if I need to lead on it, but then I don't seem to have much time either!

By Wednesday, the long-promised fine weather has settled in. How quickly the world transforms from a nightmare of muddy drudgery to a wonder-world of leafy greenness. The miracle of every spring, like the miracle of every birth, is always astounding. My little trees are unfurling their leaves, the winter-sown crops are accelerating through their growth stages, the clover is lush and the cattle are at long last released from their sheds. There is a huge amount of work to do, ploughing and drilling, moving fences twice a day for the dairy cows, who are now on overdrive with all this luscious grass, and getting the Starveall site ready for the new dairy. But spirits are lifted, and energy levels too.

The lives of our pigs and their keepers are similarly transformed. The task list suddenly shifts from bowsering water to fill frozen troughs to bowsering water to create wallows, from creating windbreaks to creating sun shades, from endless bedding-up to strimming under fences to stop the grass shorting the electricity. (Oh, for some forbidden glyphosate, the herbicide once touted as totally innocuous, but now controversially cited as a probable carcinogen, to prevent growth altogether! It does make life easier, for a whole host of jobs, including controlling pernicious weeds like couch grass, and enabling newly planted trees to get ahead of surrounding vegetation. But of course we cannot use it in organic farming, quite rightly, so we have more manual, time-consuming and expensive methods, like strimming!) The pigs' lifestyles undergo a radical change too, from burying themselves in straw for warmth to lying outside on the drying mud to stay cool. Indeed, mud is now at a premium as they compete to coat themselves in this natural sunscreen. The newborn piglets are out of their arcs in a matter of hours, nosing around in the grass, almost in danger of getting lost in it. I watch a gorgeous Saddleback sow with her 11 tiny babies. She is carefully monitoring their early adventures, calling to them and rounding them up when they stray too far. How much the weather affects our lives on the farm, and our spirits too. A few days of sunshine, and life is suddenly very wonderful.

16

High Pressure

21–30 April

It's another deliciously warm morning, and the new paddocks are full of grass. Blessed are the spring-born piglets, for they shall have sun and clover and wallows. But even Molly and her gang, born in the depths of a tough winter, now have a couple of easier months ahead of them, and they are making the most of it. Stretched out on damp mud, dozing in the late-afternoon warmth, they seem to have smiles on their faces. It is 21 April, their four-month birthday.

The lone boar piglet is still with the sows, and still unaware, it seems, that Larry and Leonard are ten times his size. Larry is chasing after 302, moving rather gingerly over the stubbly dried mud, and piglet has the temerity to get between them and grunt something of a challenge. This boy is bonkers, but he seems to be getting away with it so far.

I want to be everywhere on the farm today. The fields are buzzing with activity: muck-spreading on Fifty Acres before it is planted; rolling the new leys that will be cut for silage to stop stray stones damaging the forager knives; power-harrowing the now-ploughed remains of the turnips so that we can drill the wheat. In the morning I walk around Lower Farm, especially keen to see the Irish heifers now that they have settled in. They are looking better and much calmer too, following me back across the field but without too much frenzy. They have been moved to a newly fenced field, the Chalks, and have a wonderful clover ley to feed on, which should keep them growing quickly for the next few weeks. We want them to weigh at least 340 kg before they are introduced to the bulls in early June.

I meet up with Ben, who has brought his young family to the farm for the day to help him plant the vines. We have some trellises in the Barn Field orchard, ready to train these dessert grapes on to, and tayberries too. He takes me for a tour round the trees, pointing out details that I would miss. The almonds, so early to blossom, seem to be setting well, though it's too soon to be sure. The apricots, on the other hand, seem less happy; maybe this is not the place for them. We check the quinces, with their understorey of raspberries, and the damsons, interspersed with gooseberries. Sea buckthorn has been planted in several rows, and some of it seems to be struggling to get going, while other patches have given up the ghost altogether. Cherries have come into blossom since I was last here on Wednesday, and Ben thinks that we may even get some fruit this year, despite the youth of the trees and bushes, if we can get to it before the birds do.

With the grass growing very quickly, it's increasingly urgent to get a mulch of woodchip around the trees to stop the grass competing for light and water. A herbicide, such as the glyphosate I've mentioned before, would make this an easy job, but that's not the organic way, and so we are experimenting with using woodchip to smother the grass around the base of the trees. It's a laborious job by hand, with around 7,000 trees to protect, and the soft fruit too. Having pollarded so many willows over the winter, we now have huge piles of poor-quality wood around the fields that needs chipping, and our friendly woodland specialists are coming here soon to get this done with their big machine. We then must find as efficient a way as possible of getting the chip to the trees, and spreading it along the rows; perhaps our cattle-feeder wagon, which ejects to the side, might work? It's a valuable and indispensable part of our equipment, however, so we will want to be sure that it won't be damaged. The denser new plantings will need to be done by hand, though, perhaps with the help of a little dumper truck.

*

With so much happening on the farm, it's hard to tear myself away to catch up with emails and preparations for a busy week or two ahead, but I must. I have three speeches to give next week, plus a short-notice invitation to give evidence to the Environment, Farming and Rural Affairs Select Committee on the 'Health and Harmony' consultation. It's a lot to get ready for, and at the same time, we are trying to make decisions about the new family partnership, the Starveall dairy – still without knowing whether our landlords will support this financially in any way –

and on the new farm opportunity. We need to make progress with recruitment on the farm and within the marketing business, where we are looking for someone to help develop dairy products and the start of some online retailing. The Royal Oak financial performance is strengthening, and everyone seems to love it, but the numbers need to improve further, and quickly. Our wildlife photographer is unwell, we have just discovered, and has decided to stop his venture, so now we have six hides and a surprising number of people who would like to take on the venture – more decisions. And Claudio and his Italian crew are doing very well with their great products, and need more space to expand their processing – but where? And, of course, we need to ensure that we have a better system for the pigs next winter, especially as we have decided to keep the herd at its current size for now.

I've always thrived, or at least coped, under pressure, but this feels like the most intense period for many years. It's the same for all of us; everyone is working harder than they should be, and we need clear heads to make the decisions that are rushing at us. As ever, when life gets tough, I go to see the pigs. They always help me to put things into perspective.

And at the moment, I don't need to go very far. The orphans are on our doorstep, literally, trying to break into the house whenever the door is ajar. They are waiting by the gate when we arrive home, squealing for attention, and more importantly, food. They would eat all day now, and their trips around the garden are ever more excitable, bounding around like puppies, playing in the woods, and rooting everywhere. They really are perfectly adorable. We have developed a family routine, broken only when Tim falls ill for a few days, during which time the pigs become a perfect nuisance in his eyes. All the cute mannerisms, the squeaking, the constant need for food, the pooing in the porch, the half-eaten carrots in the yard, all suddenly become a source of irritation. I rather happily take over as Mum, luckily not for too long, both for me and for them.

There is no doubt, however, that they need rather more grown-up living arrangements. David has started to prepare the little plot by the dairy, transforming the old turkey house into a pig mansion, putting a water trough and feeder in, and surrounding the patch with an electric fence. It's very overgrown, but I have a feeling that the pigs will make short work of the vegetation. Our neighbours down there have an allotment on one edge of the field, and when they see the work in progress suggest that the pigs might like to spend some time digging over their winter brassicas, so we create a pop hole to give them access.

Their new home is nearly ready, but I am reluctant to see them go. The last few weeks have been such fun, and life won't be the same without them here.

*

This week, I seem to be running from one event to the next, speaking on a huge variety of subjects, from the opportunities for technology to help us farm more sustainably, to how we should measure the progress farms are making. Quick as a flash, though, it's the day to give evidence to the Environment and Rural Affairs Select Committee. This is a key opportunity to put forward the case for organic farming and agro-forestry as practical approaches to delivering many of the things, such as more vibrant soils, high animal welfare, clean water and carbon sequestration, that the government now suggests are high priorities. But you never know what they will ask, and so it's a moment to be on my toes.

The panel of MPs start with a question about our levels of optimism for the future of farming. It's a hard one to answer when there's so little clarity yet about what the trade and support regime will be post-Brexit! There are four of us giving evidence together, and we agree on many things, making the case that there is such an opportunity to put the public's health centre stage in the bold new world we are moving into; this means reducing pesticide and antibiotic use dramatically; growing more fruit, veg, nuts and pulses, and selling more fresh food locally so that it retains its nutrients, needs less storage and packaging, and is less expensive. It has always seemed crazy to me that fresh, unprocessed food is more costly than the ultra-processed products that are so bad for us.

Then I am asked about the potential of organic farming, the implication being that it is a minority sport that will only be of interest to a few better-off consumers. I cite the progress being made in so many other countries, where governments have given moral and financial support to organic production. In France, for instance, they have a target of 22% of food to be organic by 2020, and in Italy 15% of it already is. Schools in Copenhagen use 75% organic ingredients in school meals. This is all very possible with a bit of effort. The Soil Association's Food for Life programme is already overseeing the food in half the primary schools in England, with over a million meals served every day with some organic ingredients – though rarely as high as 75% – and with another

800,000 achieving our bronze award, so that at least the food is freshly prepared, has no nasties, and is Red Tractor-assured – the industry baseline standard for food safety – as a minimum.

One of the MPs asks several questions around the taste of organic food. I am surprised; this seems a bit of a red herring to me, even though taste is a factor that I feel is often overlooked. We all eat more fruit and veg if it's delicious, at the right level of ripeness, and prepared and presented well. Organically grown food does usually come out on top in blind tastings, but you certainly can't guarantee it; so much depends on variety, soil type, freshness and sunshine levels for plants, and breed, diet, age, sex and maturation for animals. I do know that one of the reasons many people buy organic – and this is certainly the case for our pork products – is for the taste. But I am really not sure of the relevance of this for the inquiry, and although I'm getting frustrated and keep trying to move the discussion on, knowing how short a time we have to make the big points, she keeps coming back to the issue!

Finally, to my great relief, we move on to animal welfare. We make the case for clear, honest labelling, so that consumers know what kind of life the animal has had. This has worked so well with eggs, where since there has been compulsory on-pack information, not just for higher-welfare systems like free range and organic, but for barn-reared and caged hens too, sales of free range have grown to be more than half the market. If pork was labelled in a similar way, so that people could tell whether the pig had spent its life indoors on concrete, early weaned and tail cut off, or if it had a more interesting life, on deep straw at least if kept indoors, would they choose a higher-welfare option as they have with eggs, even if it cost a few pennies more?

Then there is the question of whether animal welfare should be seen as a 'public good', that policy buzz term that describes the wider benefits from farming and food that are good for everyone, but that the farmer is not paid for. Should good animal welfare standards be mandatory, through legislation, or should farmers be paid to deliver them? Regulation is the cheapest option, but it would increase the cost of food in shops, impacting on people with low incomes most heavily. However, if the taxpayer funds welfare improvements, then the cost sits with those who pay most taxes – the better off amongst us. And, of course, the big unknown is what the standards will be for food that is imported post-Brexit. If lower than ours, then it is logical to pay farmers the increased costs they have to bear compared to our overseas competitors. So the right approach will depend on the outcome of the trading relationships and deals done with the EU and other countries. This is a more complex point to get across than a straight black-and-white answer, and I hope that I have made the complexities as clear as possible.

What we do know is that British citizens care enormously about animals, whether as companions, as wild creatures, or farmed. We make the case that if government doesn't find a way to ensure that livestock here have a good life, and if instead they allow imports of hormone-treated beef and dairy, or pork from systems that prevent any natural behaviour for the pigs, then there will be outrage – especially if these cheap and nasty imports drive our higher-welfare farmers out of business! This is an issue that cannot be left to the market to resolve, especially when it's mostly impossible to tell what you are buying. I'm trying to keep my voice level and my emotions under control. This debate matters so much, not just to farmers and consumers, but to all the animals whose quality of life will be determined by the outcome. Around ten million pigs are produced in the UK each year; their prospects hang in the Brexit balance. After an hour and a half of intensive questioning, the session ends. There's a chance for a few more informal words with a couple of the MPs, and a short debrief between the four of us, then that strange flat feeling as the adrenaline seeps away, and I wend my way home.

*

I grab a few minutes early the next morning to see what the pigs think about it all. Molly and co. are not especially talkative initially; maybe it's a bit too early for them. I leave them to wake up, and wander off to see the newly farrowed sows. In the cool of the morning, a particularly delightful Saddleback litter are exploring the world under the careful supervision of their mother. She is chatting to them slightly anxiously, rounding them up when they stray too far from the arc, and nudging them back towards home. I could watch them all day; they are just so enchanting, and as I do, the debate from the previous day plays back in my head. Far away from the stuffy air of the committee room, out here with my flesh-and-blood pigs, I am as ever reinvigorated and reinspired by them and by this land that I love.

David arrives in the field, and is pleased to tell me he has been catching up with pregnancy testing, and that six of our eight sows, including Mrs Duroc and Molly's mum, are back in pig, though not, as yet, Mrs Messy Bed. Pregnancy testing is a relatively easy task if everyone behaves. He feeds the sows, then moves between them with the doppler scanner, running it against their sides to pick up a picture of their uterus, very like a human ultrasound scan. Ideally, we do this around four to five weeks after the boars have gone in, and again at eight weeks or so, to ensure that the sow has not reabsorbed her litter. If all remains well, this means that most of the sows will farrow again in mid-July, a month later than it should have been because of the late weaning of their previous litter.

I go to congratulate the sows. Messy Bed is still hanging out with one of the boars, perhaps aware that getting pregnant is a key requirement if you want a long life on a pig farm, but more likely because she seems to prefer the company of boars to those of the same sex, or even of her piglets. The others all come up for a scratch, however, sleek and content. Then I track back to their piglets to see if they have woken up yet, and this time I'm mobbed by the curious creatures. Their enthusiasm is rather overwhelming these days, in that they are so strong now that they can almost knock you over. They must weigh around 60 kg, and some of them are big enough to qualify as 'porkers'. Traditionally, fresh pork comes from smaller pigs, while bacon and sausages are produced from larger 'baconers', around 110 kg in weight. As much of our produce is, indeed, bacon and sausages, our pigs will be slaughtered at the heavier end of the spectrum.

Delayed by all this attention, I nearly miss my train to London. Another day, another conference; it seems the whole farming and food world is constantly conferring with so much at stake. When I get home late in the evening, I am immediately struck by the silence and the open yard gate. It takes me a bewildered moment to work out what has changed, and then the penny drops: our orphans have gone. Will life ever be the same again? No noisy greeting, no poo on the doorstep, no trying to squeeze through the door without letting them into the house. They were only with us for three eventful months, but they have wormed their way into our lives and hearts, and the place feels too quiet, too empty, without them.

Of course, they haven't moved very far, just to their much more appropriate home in the turkey mansion. Tim gets back, and tells me that they trotted quite happily on to the little trailer that David brought round, and seem entirely unphased by their changed circumstances. But he is worried that they may not have found the water trough, or worked out how the feeder operates, so he is off again, to check on his charges before night falls.

Back for the weekend, I go to see them myself. When I arrive at their field, just one call and across they scamper, rubbing up against my legs like rather ungainly cats, trying to work out if I have anything good to eat. They take me on a tour of the facilities. There's a path from the water trough to the house, a nice patch of nettles to the south of the field, much overgrown grass in the west, and that wonderful pop hole into the neighbours' veg patch. Yes, that's a real bonus, they say. I find that Tim has installed two chairs by the doorway to their house, and I can imagine that he is spending many of his non-working minutes sitting here having a Hamlet moment. Without doubt, these pigs have landed on their trotters.

*

Much of the rest of the weekend is spent trying to make decisions with Henry, Sophie and Dai. We still haven't had any news from our landlords as to whether they are willing to invest with us in the new dairy, or even to accept any investment we make as a tenant's improvement. But if we are to get the dairy built by next spring, we need to confirm the builder, who is coming from New Zealand, and pay a substantial deposit. We have already bought the Irish heifers and paid a deposit on the equipment; all this is just about reversible – we could sell the heifers, and even the equipment – but once we have paid the builder, we really are at a point of no return. This is such a huge investment, which will increase our costs and, of course, our revenue too for many decades to come, however we finance it. I am sure it is the right thing to do, given Sophie and Dai's passion for cows, but it sets us on a new course, and commits our farming family to another lifetime with cows. I am both daunted and exhilarated, feeling the pressure of making these pivotal decisions with too little information, but with no time to get more.

We also need to buy enough bulls to serve the heifers, and soon enough that they can be quarantined for a month before they get to work. There's also the question of whether or not we should buy Friesian bulls, if we can find them, or beef ones, which are quieter to handle, but will mean we have fewer females to replace any we lose if we get TB again.

Henry and Dai are somewhat exhausted by managing the new land, with all its emerging complications of staff who are reluctant to try new ideas, fields that should have been planted with grass last year and haven't been, and the need to get animals out on to pasture that is still only growing slowly. We must work out the best long-term arrangements for us and the farm, and we tend to veer from thinking that this is a wonderful opportunity that allows us to rear all the calves we produce, giving them the sort of upbringing they deserve, and to keep all the ewes in one place, to feeling that it is a complication too far at such a time of change and investment at Eastbrook.

Despite their heavy workloads, what is clear is that Henry and Dai have been doing a brilliant job in getting things moving on the new estate. All the spring cropping is drilled, and quite a lot of the undersowing of grass too; the cows and calves are out, machinery is getting mended, and the grain store is at least partly cleaned. At Eastbrook too, even though the land is heavier, and therefore slower to dry, we are well underway with planting – despite this all happening six weeks later than it should have been – and all animals who can be out now, are.

By Sunday we have made one or two decisions. Dai has found some Montbeliarde bulls in Cumbria, and we agree that this is a perfect compromise. Montbeliardes are a 'dual purpose' breed, which means that they will produce reasonable amounts of milk if we need them for the dairy, but also make splendid beef animals if we don't. If our vet is happy, we will buy them. In an ideal world we would go to see them on the farm, but no one has time for a trip to Cumbria, so we are looking at photos and relying on the assurances of the very pleasant family who are selling them that they are in good shape.

We have also agreed that we must assume we will need to fund the dairy ourselves. We cannot afford to do the new shed, so somehow we will need to make do with the old one for a while if our landlords decide not to help. But we will need to borrow much more than I had hoped, and we go through the figures over and over again, trying to make sure that they are not too optimistic, running through all the 'what if?' scenarios. We are meeting the bank manager next week; he is new to us, even though we have been banking with the same business for decades. What will he make of our plans?

Finally, we pack up, and Sophie heads up to Nottingham. She starts her veterinary exams in the morning, and having been here for the last month, helping with lambing and much else, I'm not sure when she has found the time to revise. She seems calm enough, however, but then Sophie always is.

*

I head for the pig field. The good weather has vanished, and this evening it is cold, and spitting rain. I leave Dog in the truck and set off unencumbered across the paddocks. The pigs hate dogs, so it's much easier to have a good look at them if I don't have him in tow.

There are several new litters, all tucked up in the warm. The teenage mum is out and about though, with her seven gorgeous piglets, and is much less protective now that they are ten days old. I walk across to Molly and her friends, and I'm pleased to see that the team have put a bale of silage into their paddocks. With this wet spring, the grass is now almost entirely destroyed, but they still love to have something green to eat. Silage is a good substitute for fresh grass, and the growing pigs are both chomping and lying on it.

Having chatted to the girls, I go across the fence to the boys. There has been a noticeable difference between the sexes for a few weeks, with the boars always rather less friendly, and now I can hear a difference too. A deeper grunting, from one or two especially, as though their voices have broken. I keep moving, walking through all the older growing-pig paddocks. These pens have three large arcs in them now to accommodate the size of the pigs, and their inhabitants charge out as I approach, chasing around the arcs, bouncing around on the straw, full of energy and playfulness.

Beyond them, the field is mostly cleared now of arcs, fencing and troughs, and the straw from the arcs and the outdoor pads has been heaped up ready for spreading. The tracks have been cultivated, and this part of the field will soon be ready to plant with turnips. What looked like a bomb site a couple of weeks ago after the wet winter is starting to feel like a normal field again.

I walk on, across Starveall Penning, now rolled and ready for silage making in a few weeks' time, and into Starveall Paddock, where the ewes expecting triplets have been lambing. The lambs are at their cutest stage, small but plump, getting together in gangs as they race around, jumping on and off the straw bales that had been put there for a bit of shelter during the worst of the weather. Three lambs are curled up under the hedge, looking as though butter wouldn't melt in their mouths. The lambs too have recovered from their difficult start in life, and are enjoying their time in the sun.

I track back across the arable crops in Haybarn and Sixty Acres. They are growing fast, making up for lost time, fuelled by the fertility that the pigs left behind when they were here three years ago. The crops are free from disease, and with very few weeds too; the pigs do a great job of cleaning our fields of docks and thistles, and the strong, lush crops that flourish after them out-compete most of the rest. I like some of the right kinds of weeds, plants that will sustain the ladybirds and predator wasps – insects that will come to the rescue when aphids attack – but like all farmers, I love to see a clean crop. It's bred into us! With the rain falling more heavily, my legs are soon soaked as I stride through them, and when I get back to the pigs, I take shelter under the tall hedge, and watch for a while. Despite the cold rain, many of the older pigs, including Molly's gang, are scattered across their paddocks, avidly rooting. This, as a very minimum standard of care, is how we should let our pigs live. There is more we could do to make their lives even better, and we will continue to strive for this; but all the basics that make for a good life for pigs are here: soil, and straw, and freedom. All pigs deserve this kind of life. It's a cause worth fighting for as long as we continue to eat them.

After a week of dry weather, the rain is softening the ground again, bringing worms to the surface, and allowing the pigs to dig deeply. They certainly seem to be making the most of the endless nirvana that the soil provides.

17

May Day

It's a big week for us as the month of May begins. The deadline for making a final decision about the new dairy and whether we take on the land that has been offered is upon us. We still don't know if our landlords want to help financially, and if so on what terms. The New Zealand builder also wants to come here as soon as he arrives in the UK in July, which is great in many ways, but we will lose him if we don't pay the deposit today. We have been prevaricating for several weeks, so on May Day, it is D-Day.

The pigs seem to have no idea of my turmoil when I visit them early in the morning. They crowd and boot-bite relentlessly until I retreat to chat from behind the security of the electric fence, at which point they wander off to find other attractions. While the field of pigs goes about its daily routines, I park myself on a bale of straw, and try to work through our options, with only birdsong and the distant rumbling of tractors disturbing the peace.

I spend the morning taking advice, and talking with our landlords' agents, and juggling calls with my Soil Association team too. Another team of Defra officials are coming for the day at 11 a.m. to learn more about farming as they grapple with translating the ideas that Gove has been outlining into workable solutions for farming and the countryside – so I need to have made a decision before they arrive. With a couple of the jigsaw pieces starting to fall into place, we take the leap and put the payment through. There's still some risk, but I know in my bones that somehow we will find a way to finance this. It's a game-changing and nerve-racking moment, but there's no time to reflect on it as I get caught up with my visitors. It's a bright day, and I need to walk, to calm and focus myself, so I drag them through pigs and crops and grassland, into the Valley too, to see the carpet of cowslips and wildflowers that now adorns its banks.

Over a well-earned lunch we talk about resilience, one of the themes that the officials are working on. With volatile markets, and increasingly volatile weather too, government is aware that farmers will struggle to cope with the elements outside of their control – and this will be even harder to do without the safety net of the farm payments we currently get from Brussels. They are interested in our approach, and the efforts we have made to spread risk by having several enterprises, and by adding value to at least some of what we produce.

Although it looks successful from the outside, I'm quick to warn that it's been a long, hard road, with many 'skin of our teeth' moments along the way. We made mistakes early on, mainly through not having good enough financial information and advice, but as I tell my guests, it was the foot-and-mouth outbreaks that nearly finished us, twice.

It's easy to forget how devastating the 2001 outbreak was. I was a Meat and Livestock Commissioner at the time, and we had a meeting on the day that the disease was confirmed in the UK. As we gathered, one of the most experienced and powerful operators in the meat world said to me: 'It's the farmers who don't get it who will be put out of business,' and by the end of the epidemic, I understood this all too well.

At the time, we were working with around 30 other pig farmers as a result of a joint venture we had set up with a large pig and feed business who wanted help to develop the organic side of their operation. They had abattoirs and processing plants, and we had the knowledge of organic farming, plus a small gang of producers we already worked with. It seemed like a marriage made in heaven, except that six months in, the parent company sold the pig business, leaving us with none of the manufacturing assets that had made this relationship attractive. To cut a long story short, we went our separate ways, but by then we had many farmers contracted to supply us; quite a responsibility for a small enterprise like ours.

Two weeks into the outbreak, my then pig manager turned up on the doorstep early in the morning, as white as a ghost. He had spotted some lame pigs, and thought there might be suspicious lesions too. We called the government agency dealing with suspected outbreaks, and immediately all hell broke loose. Within an hour, a vet was on the farm, and I spent the day – a wet, cold, windy day – checking every group with him. He thought it was a false alarm, but sent samples away for testing. Until the samples had been analysed, we were on lockdown. No vehicles in or out, no people in or out. Sophie was at school a mile away, and was not allowed home. She was 12 years old, and as traumatized as I've ever known her to be at the thought that all our livestock might be slaughtered.

Thankfully for us, it was a false alarm, but it took several agonizing days to find that out. Very soon, abattoirs were suspended if they were in an infected area, and many of our farmers could not move their pigs to slaughter. A backlog of pigs was building up on many farms, including our own, and as the weeks and months went by, this was in danger of becoming a welfare problem. We couldn't even move pigs to another field if that meant they had to cross a road, so they were trapped in muddy fields, growing heavier every day.

By the end of the outbreak, we had thousands of pigs on farms that needed to be slaughtered all at once, and no idea how we were going to turn them into cash. We had been unable to supply the supermarkets with our products during the crisis, and now we had far more meat than we could deal with. We were contracted to buy the pigs, and we did so, freezing most of the meat until we could work out what to do with it. We had no cash left at all, just a frozen pork mountain that nobody wanted.

One of the group asks me if we considered giving up at this point, and I tell them that it looked for a while as though we might have no choice. The lowest moment came when we took advice from an insolvency practitioner. His counsel was clear: go into receivership, get rid of your debts to the farmers and others, start again if you want to, using a 'pre-pack' arrangement. It was sound advice in the circumstances, but it made me angry. I had always been cynical at the way companies could just ditch their debts, with all the impact that has on their suppliers, then change the company name by a couple of syllables, and carry on as if nothing had happened. All our farmers' livelihoods would be threatened if we took this course, and I wasn't sure how we could live with ourselves.

We decided we should fight on. We met with our farmer suppliers, and talked them through the situation. They backed us to keep going, agreed delayed payments, and cooperated magnificently with us and each other. Now we just had to find a way to turn it around, to do something with the pork mountain.

Tim and I went to Germany. We had been impressed on our visits there to trade shows by the expertise and fabulous pork products that the Germans have. Pork is their favourite meat, and they are so much more inventive with it than we are. We had especially loved the small sausages that are a speciality of the Nuremberg region, nearly 100% meat, unlike the British banger that is bulked out with rusk, and which they eat with potatoes and sauerkraut. We had done a little business with a family company who made them, and we went to see them, to ask them to make them for us.

One of the many great things about this sausage is that it has a much longer shelf life than a normal one. We were already selling sausages into the supermarkets, but as we cannot use artificial preservatives in organic food, the shelf life was only six days, compared to a preservative-enriched product, which would have ten or 12 days. This meant that we were always being charged for wastage by the supermarket, and once this had been factored in, we made no money at all on the sausages. But the little 'speedy', as we named it, was pasteurized, so it was safe to eat for over a month, and quick to cook – hence the name – without risk of food poisoning.

We launched it into Sainsbury's in 2002 with our orange pig riding a motorbike on the packaging to underline the rapid-cook angle. Sales started to climb, despite the fact that we couldn't afford to promote the product as we should have done. It was young families who were driving sales; the idea of a 96% meat, organic sausage that cooked in five minutes seemed the perfect solution for time-strapped parents. It is still a best-seller, 16 years later.

We launched other products too, some successful, others not, and built a strong relationship with our German partner, supplying them with fresh pork for their organic products too. Gradually, the pork mountain became a pork hill, and we were able to start to clear our debts. By 2007 we could see the light. But disaster struck for a second time as another foot-and-mouth outbreak hit our countryside.

Many people will have forgotten the second wave of foot-and-mouth. It was nowhere near as traumatic as the first; many lessons had been learned, including one that the Soil Association had campaigned on vigorously the first time around: the use of vaccination, rather than preventative culling, to control the spread of the disease. This time, there were no searing pictures of pyres of animals burning, the countryside did not grind to a halt, but the export of meat was stopped, for all the logical reasons. Over half the pork we produced every week was now going to Germany so this time we had a huge problem; lots of pork with no home, and no speedies to keep our customers and bank manager happy.

It was another setback, one which drove us to launch a range of new products with the UK retailers, notably with Tesco, and to start to work with some Swedish pig farmers who were prepared to upgrade their standards to the higher ones required by the Soil Association. We needed to know that if this happened again, we had other options for supply into Germany. But of course, you know what happens next – the financial crash of 2008, which saw, amongst many other things, the more cost-conscious supermarkets like Tesco cut their lines of organic food dramatically. They just assumed that their customers would not want to pay any more than they had to during the recession. And so, our carefully developed organic pork and veal products were in store for just three weeks before all of the cost and effort went down the drain.

The more far-sighted Sainsbury's stuck with us, and we developed other markets too, with online retailer Ocado, and with box schemes like Riverford, and Able and Cole. Again, we hauled our way back to solvency, but goodness me, it's not been easy. We have done, and still do, many things that in a perfect world would not be necessary, because attempting to make a living in farming and food is not straightforward, especially if you are trying to do it ethically. I think we are well placed to talk about resilience, I tell my Defra guests, and hope that sharing some of this turgid tale helps to dispel any sense that what may look like success at Eastbrook today is easily replicated.

'What would make it easier?' one of Defra officials asks.

I want to bring Tim in on this bit of the conversation. He was the one running the marketing business through these turbulent times, and he has strong views on the lunacy of the way we run our food supply systems, but he is fully occupied serving customers, so I plough on. One big opportunity now, as I see it, is to invest in local and regional infrastructure, like abattoirs, dairies, fruit and vegetable processing units, and hubs to store and despatch, so that farmers and growers can get their food to customers quickly and efficiently, without a million middlemen who all need to take a margin. We end up wasting loads of food because it doesn't reach the retail specification, or because they've over-ordered; we over-process it to add shelf life and margin, then we over-package it to protect it through all the transport and storage it has to go through, so by the time it gets to the customer, it's either not very affordable, or not very healthy, or both. And despite all of this, most farmers will still be getting less than the cost of production; no wonder they are so worried about losing the EU support payments.

Ben arrives and takes them off to see the agro-forestry in the hope that this will help them see what potential there is in supporting tree crops as part of future agricultural policy, while I go to the orphans, to see if they would like a banana or an egg. I follow Tim's advice that the eggs go in whole now, like some ridiculous party trick; one crunch, then lots of dribbling as they lick the insides out and reject the shell. They take me to see their allotment, where we find the neighbour's dog, a new friend for them to taunt. They chase it around, grab at its collar and tail, a daft game which the dog seems pleased to indulge in. There is evidence of their rooting in several places, but with so much room, it's a spasmodic rather than systematic clearing of this overgrown patch. They are still operating as a twosome in all that they do. They give the electric fence a wide berth, and have clearly made themselves very much at home in this piggy paradise.

Speaking of animals making themselves at home in new environments, on Thursday, six rather splendid Montbeliarde bulls arrive in a lorry the size of a house. I happen to be at Lower Farm when they turn up, so I get involved in the unloading of these magnificent youngsters, who are very keen to get out on to some grass after their long journey. Even so, they baulk at the lorry ramp when the tailgate is lowered; animals are always so wary of moving across different surfaces. Our orphan pigs were just the same in the garden, taking a while to adjust to gravel, or snow, or wooden decking (although they never seemed nervous of my kitchen floor, or carpets, I seem to recall!). I'm about to go to find some straw when they leap to freedom, landing in the still rather mushy gateway, and haring off around the field, bucking and kicking. It's definitely time to shut the gate! I turn my back for a minute to secure the field and when I look into the pasture I see that the heifers a couple of fields away have all lined up along the fence as if to say: 'Hello there, sailors!' to the new bulls in town. The attention the heifers are giving them has not gone unnoticed by the bulls, and I'm glad that there's a stream between them. I have a feeling that testosterone would overcome most fences.

*

The next morning finds me up extra early, having slept badly. My head is full of the day's crucial meetings, first with the owners and agent of the land we are looking after, and then with the bank to try to secure the borrowing we need to go ahead with the dairy build, regardless of our landlords' decision on investment. I walk up to the Valley to clear my mind, past the yearling cattle on Lower Nell who are starting to look glossy with all this spring grass inside them. The cowslips in the Valley are wondrous now, and over the last few years have spread from the banks, where the fertility is poorer, on to the valley bottom and the flat areas above the slopes. Without the competition of nitrogen-loving ryegrasses they and many other species can thrive more readily. Since we have been grazing it less heavily through mid-season, more hawthorn and blackthorn are becoming established on the banks too, providing additional feed for birds and shelter for mammals. It is one of my favourite places. As children, we played and picnicked in the hummocky mounds near to the big badger sett, and as teenagers my brother and I caught rabbits here with his ferrets. As it is so close to the farmhouse, I still walk here often. From the top you can see for miles, and the steep climb is worth it, both for the view and the chance to properly stretch my legs.

By evening, it feels as if it's been a very good day. We have agreed with the estate that we will take the grassland there on a grazing licence, and they will find someone else to do the arable cropping; we thought about doing both, mainly because we want to make sure that the whole rotation works well and improves the soil, but we really don't want to be buying more arable equipment and taking on all the work it entails at the moment. And the new bank manager turns out to be a charming man, who started the conversation by saying that our financial reporting to him across all the businesses was better than he gets from any of his other clients. I guess largely due to that, he seemed to understand what we are trying to do, and helpfully suggested a number of ways to make our situation as flexible as possible. Without any hesitation, he offered us what we need, which means we can keep moving forward at pace.

It's the beginning of the spring bank holiday weekend, with the weather set fair, and a couple of weights off my mind. To celebrate, Tim and I take a beer to the orphans, for us not them, and sit on the wonky chairs in the sunshine with our pigs snuffling around in the undergrowth. We reflect on the day, another important moment in the ongoing saga of our rural enterprises, trying to survive and thrive through all the challenges over the decades. Our resilience has taken many forms: diversity in what we do, finding great people to help us, taking calculated risks, and adapting quickly when needs be. And more than a stroke of luck too, on occasion. Just for a moment, beer in hand and the soft grunts of the pigs in the background, I think to myself that it has all been worth it, that life doesn't get much better than this.

18

Sunshine and Shadows

13 May

The sunshine has put a spring in everybody's step. Grass has never grown so quickly on this farm, making up for lost time after the revoltingly cold start to the season, now itself almost a distant memory. The dairy cows are in heaven in their succulent pastures, and rewarding us with oceans of milk. Teo stops me this morning to say that we are selling over 5,000 litres a day, which I think is the most that we have ever produced. Silage making will be upon us in a flash: how fast the year spins round.

Molly and her gang are growing at a rate of knots too, though the hot weather leaves them lethargic through the middle of the day, and they are more interested in mud than eating. They have a wallow now, a pit that we dig, putting the soil carefully to one side behind an electric wire so that it can be replaced once they leave, and then fill with water from the bowser. The instant paddling pool soon becomes a gloopy, slimy mud bath, filled with immersed pigs that could be mistaken for baby hippos. Where once they fought to be in the centre of the deepest pile of straw, now they compete to be in the deepest part of the wallow. They emerge caked in mud which slowly dries like a face mask and protects them from the sun.

In the early morning they are still lively, however, and go through their ritual of mock horror when I arrive before overwhelming me with attention. Molly, like her mother, is more aloof than most of them; she hangs back, curious but wary, while the Saddleback twins who were the cuckoos in her litter are bold and naughty, barging through my legs and almost knocking me over.

The sows and boars are at their most active at this time of day too. We have started the three-week batch approach, and last week David weaned nearly 200 piglets from 22 sows. He needs to serve around 30 this week, to ensure that 24 farrow together in four months' time, so he has added some maiden gilts into the herd, and a few sows who failed to become pregnant last time around. This is a busy week for the boars, and they are hard at work, frothing at the mouth, tracking the sows that are coming into estrous, snapping at their colleagues who offer to come and help.

When David arrives, his first words are 'Thank you for giving me Rachel!' He is feeling much more on top of the workload now, and has even managed to spend a day putting metal stakes deep into the ground to ensure that the electric fences have a strong earth. This is the kind of job it's easy to delay when there's too much to do but, especially with the drying ground, it will pay dividends in making sure that the pigs stay where they are supposed to. As I walk round with him, I can hear a lively clicking noise wherever the soil is touching the wires, a good sign that plenty of current is going through. The next job I want him to get on with is to start to use the whey from the mad Italians better; we are wasting too much, and need to introduce it to the piglets as soon as they are weaned so that they get a taste for it early on in their lives.

It's only three weeks until pig racing, our crazy day at the pub, so today he will select the favoured eight for their moment in the spotlight. These need to be pigs that are at the right age, around ten weeks, so that they are confident and strong, but not a danger to small children! And they need to be readily identifiable by the punters. There is a litter David has had his eye on since birth, a lovely hotch-potch of colours and markings, and we go to see them. I agree that they look perfect, and that he should bring them down to the farmyard so they can begin their 'training'.

*

I walk down to the Forty buildings, near the dairy on the edge of the village, where the pigs come for their last few days before slaughter. There are four deep-straw pens here, each with an outdoor area, a concrete strip with feed hoppers and water troughs. This allows the pigs to get used to a hard surface after their life on soil, and provides a space for us to weigh and sort them. The whole place is designed to make handling the pigs as stress-free as possible, both for them and us. The outdoor run curves around at the far end on to a gently rising loading ramp for the lorries to pull up to. Again, we want to avoid any changes in surface, and pigs don't like metal ramps, so the system gets them to the right level before they enter the lorry.

Despite the space and bedding, the pigs make much more of a racket in this building than they ever do in the field. Most of the time, all is peaceful, but by this stage of their lives, the boars are constantly horny, and spend much of their time trying to ride each other. It's a rather under-discussed area of pig welfare. Very few pigs are castrated in the UK, and that's great in many ways; I'm all in favour of avoiding mutilations wherever possible. But when you see a boar screeching away as another tries to have his wicked way with him, you do wonder. We keep the gilts separate from the boars after weaning for precisely this reason – and of course, we don't want them to become pregnant – but in the absence of females, and with hormones raging, the boys prey on each other rather mercilessly. It happens in the field too, but it's easier to escape there, and there's more to keep them occupied too.

Tony is down here sorting the pigs that will leave this afternoon. There are 50 to go, and he selects the heaviest from the boar and the gilt pens, keeping them separate; the last thing we want at this stage is groups getting together as they will scrap to establish dominance. Calm is the order of the day.

The lorry arrives at 2 p.m., and the two groups load without much fuss, thanks to plenty of straw on the ramp, which helps enormously. Less than 30 minutes after he arrived, our trusty driver of nearly 20 years is away. It's only an hour from here to the abattoir, though to be honest, once they are on the lorry they settle down quickly; it's the loading and unloading that is the most potentially stressful time. They go into lairage pens overnight, with some organic feed available on arrival, so that they can be first down the slaughter line in the morning. Organic animals must be killed on a clean line, so there is no risk of cross-contamination by blood from other stock, and therefore most abattoirs will deal with organic animals as the first job of the day.

This is the hardest part of our lives with pigs, and I'm very aware today that it won't be many more weeks before Molly and her gang are starting their final journey. I wonder how I will feel about this, given how close I have been with them over the last few months. I used to have a horror of abattoirs, but then I spent a lot of time in a whole range of them, both here and overseas, and strangely enough, seeing how well death can be managed took much of the abhorrence away. We project much of our own terror of death on to the idea of killing animals, but, done well, the end of life for livestock can be a great deal less traumatic than the prolonged process we put members of our own species through.

Our pigs are killed by gas stunning. There are advantages and disadvantages to this compared to the main alternative, electrical stunning. Using electric tongs on the head to stun pigs knocks them out instantaneously as long as they are properly positioned, but they need to be dealt with one by one with their pen mates around them, or separated out, which they don't like at all, and it's inevitable that the operator causes some stress as he moves through the group. With gas, they stay together throughout, lowered with their pen mates into the tank. There is an aversive reaction for perhaps 10 to 15 seconds. Having seen both systems over the years, I would choose the gas. In a well-run plant, the pigs stay very calm until those final moments. It's still not perfect, though, and I am interested in some of the new methods that are being developed, especially something called Low Atmosphere Pressure Stunning, which has been approved for poultry recently, and is now being investigated for use with pigs. Having cared for our animals all their lives, we all want their deaths to be as calm and painless as possible.

Once the pigs are dead they are then bled, the carcasses scalded to remove the hair, eviscerated and left to cool in fridges overnight. The next evening, the 150 or so carcasses, from our farm and from several others that we work with, are loaded on to a lorry and leave before dawn to arrive at the cutting plant by 5 a.m. each Friday. A team of butchers will work all morning, breaking the carcasses down into 'primals', the main chunks of meat, such as legs, loins, belly and shoulders. Our sales team will have given the precise details, the 'cutting list', of the specifications for each primal, depending on the final destination of the meat. So for loins that are to be used for bacon, there will be instructions on how much fat cover, whether rind on or off, the minimum and maximum length and width of the loin, how it should be packed and labelled, who will collect it, at what time and where it is going. The meat must stay under 5°C, so it's cold work, and the butchers will be wrapped up warm in clothing that protects against both the chill and the machines they use to cut and pack.

The great thing about pork is, as the saying goes, you can use everything apart from the squeak. Shoulder and trimmings go into our sausages, the legs for ham, the loins and bellies for bacon, pork chops or belly pork, a perennial favourite in the pub. The lungs and livers and ears and trotters find a home though mostly as pet food as, unlike some other nations, UK consumers are not so keen on the offal and trotters. Even the skin has a good market, mostly used for collagen for the pharmaceutical industry. By mid-afternoon, the meat is cut, packed and labelled, ready for the next stage of its journey to our tables.

Bacon is the mainstay of our branded business, with two manufacturers curing it for us, one in Suffolk, who makes smoked bacon for Sainsbury's and Riverford, and one in Northamptonshire, who makes a 'green', unsmoked bacon that sells through Ocado. Both use a dry-curing technique, the best and most traditional way to make bacon. The curing salts, a mix of salt and saltpetre, are rubbed by hand into the loins and bellies, and then the meat is stacked for ten days, allowing the salts to penetrate, and the excess liquid to drain. This is in stark contrast to the way mass-market bacon is produced, where triphosphates are used to increase the water content of the bacon. Our bacon will be only 95% of the weight of the meat at the start of the cure, while many standard bacons will weigh 20% more. It's one reason why our bacon is more expensive – the customer is buying 25% more meat – and a further advantage of this drier bacon is that there is far less oozing on cooking. It tastes much better too, or so our customers continually tell me. If we are going to eat animals, then at the very least we should appreciate and enjoy them as much as possible.

It's such a responsibility, the lives and deaths of all these sentient beings. I can understand those who want nothing to do with it, and yet whatever we eat, however we live, we have blood on our hands. We chop down the rainforests to grow palm oil and soya, we destroy thousands of creatures and their habitats whenever we build a house or a road, we poison rats and mice – a ghastly death – and kill animals on the highway with our cars. And you don't have to look any further than your favourite David Attenborough programme to understand that in nature, everything is killing something else all the time. We are part of that nature, whether we like it or not. But we have a duty to tread more carefully on the earth. Even if we accept our self-assumed role as top predator, and have decided to domesticate animals rather than hunt them, we still need to understand the implications of our addiction to meat. Around a third of the crops we produce are fed to livestock, enough to feed another 3.5 billion people if we ate the crops rather than the animals.

*

Having waved the lorry off, I go to see the eight multicoloured youngsters who have been selected for pig racing. They are rather bemused by their new surroundings, climbing over mounds of straw, and looking wary of their human visitor. From previous experience, I know it won't be long before they are enjoying the limelight, and all the attention that they will get from the office staff next door.

When I go and find the orphans later, they are stretched out in the shade of the allotment. They have so fallen on their feet, these two, with their own private wallow, a veg patch to dig over, endless visitors bearing gifts, usually bananas, and more space than they know what to do with. I wonder whether they would put up with the racing pigs as companions; the orphans have been on their own for so long now that I have no idea whether they would welcome or resent other pigs, but it may be worth a try.

I have neglected the land at the bottom end of the farm recently, with all my focus on the pigs, and I especially want to see the spring-born calves, who have now been turned out with their foster-mums on to a glorious clover ley in the field we call Pidmer. I take Dog with me, as it's not long walk from the road; he is still keen to go for walks, but after half a mile he is tired, and wants to turn for home.

The calves are all together under the willows near the gateway to the field, with three cows on supervision duty. Way across the field, I see the rest of the cows, grazing in the evening sunshine, with just a smattering of older calves, who are clearly deemed mature enough to leave the nursery. They head in our direction as they see us approach, keen to check that Dog poses no threat to the babies they have left behind. They walk sedately over, while the calves with them race around, bucking and weaving. The cows start to call to their calves back in the nursery, who get to their feet and move hesitantly towards the herd. As they come together there is much nuzzling as they find their charges, and the calves start to feed. This is a very lovely way to rear the next generation; to be able to have them outside on grass from a week or two of age, behaving as a herd.

The days are long now, and there is still time to visit the pigs before dark. The sows, however, have already turned in for the night, except for one of the gilts. Six slumbering giants cover the floor of one big arc, with a small patch unoccupied near the entrance. The gilt 312 makes a move towards the space, and in unison a chorus of warning gruntles begins. She stops, ponders, puts one foot tentatively in; she stands there until the wave of displeasure subsides, then edges another leg forward. Again, the warnings start, and again she waits. Once half her body is through, and she is beyond the point of no return, she sidles in and settles down, ignoring the residual outrage. There's much shuffling in semi-sleep, small adjustments that make enough room, and now a breathing carpet of pig bodies fills the arc with relaxed snoring.

There is one sow missing, of course. The smaller arc next door has a single sow in it, stretched out in splendour. Mrs Messy Bed is, yet again, hogging all the duvet.

19

Make Hay While the Sun Shines

15–30 May

It's 15 May, a cool morning with the promise of warm sun later. Tony is busy filling the wallows and mending water leaks and overflowing troughs. This is a perfect morning for pigs. Molly and her pen-mates are lying around on fresh straw, or nosing away at the remnants of their silage bale. All the sows are still in bed, except for Mrs Duroc and Messy Bed, who are out and about making the most of the fresh morning air.

The weather forecast is set fair for the next ten days, so although it's a week or two earlier than usual, we decide to try to get the first cut of silage made. This is the most critical moment of the year, as the quality of the mountain of preserved grass we will make determines how the cows will milk all winter, and whether they will need much supplementary feed in the form of grains and protein crops, like pulses and soya. There is such a judgement call to make between the volume and the quality: the longer we leave the crop, the more of it there will be, but the nutritional value declines as the grass ages. And then, of course, there's that forever unpredictable factor, the weather. If the cut grass and clover is rained on, it will deteriorate, and to get the best quality, we want to cut while the sun is shining, when the sugars in the crop are high – this is vital both for the feed value, and to help it pickle well.

It's all hands on deck for this job too. We need at least seven people to enable us to fill the clamp as quickly as possible; one to mow, one to row up, one to fork it into the clamp, three on trailers, and one on the big forager, which chops the grass and jets it into the trailers. With our long, thin strip of land, we can be moving grass a couple of miles, from the fields at the bottom end of the farm to the clamps – two big chalk and railway-sleeper-sided pits – just below the Ridgeway. At the morning meeting, Andy calls in sick, so Sophie offers to drive one of the trailers. Our most experienced tractor operator, Andrew, puts her through her paces, making sure she is confident with the ten-tonne loads she will be manoeuvring from field to clamp.

Towards the end of the day, I get the chance to go to see how they are getting on, and to grab a moment with Henry, Sophie and Dai. Our landlords are finally coming to the farm on Friday, to talk through the dairy project, and as the boys can't be spared from the silage gang, Sophie and I may have to lead this important conversation without them. I'm keen to take their views, though, and as the tractors stop for the night, we lean against the trailers in the farmyard, and chat it all through.

Dog seems to have regained a bit of energy over the last few days, which I wonder might be due to those pesky pigs getting out of his hair. So, talking finished – everyone's too tired to go on for long – we head off for our favourite walk over the Downs.

As we set off up the 'bumpy track' I see that Branston and Pickle, the two new Hereford bulls, have been joined by the four Anguses. It was perhaps a bit brave to mix them, knowing the propensity of unfamiliar males of all species to fight with each other, but they seem to have settled down well together, with a few barren ewes for company too. These were shorn yesterday, another huge job in the spring calendar, which it's vital to get done before the flies decide to lay their eggs in the fleeces. Fly strike, as it's known, is one of the most unpleasant welfare problems to afflict sheep; the maggots eat into the flesh, and if not spotted quickly, can kill the animal. Every day, we will check carefully for any dark patches in the wool, or sheep that are irritated; you can spot the signs quickly once you know what to look for.

Dog still seems happy when we get to the top, so we head along the ridge, alongside our neighbours' rapidly growing, spotless crops, on to our grassy downland. There is a big mob of cattle grazing in rotation across these old pastures, and where the grass is regrowing, the yellow of the dandelions that were flowering in such profusion a few weeks ago has given way to the yellow of buttercups. Most farmers would think of both as weeds, but the cattle seem to eat them as enthusiastically as they do the clover and are always so content on these swards. I don't even mind the odd patch of nettles; once cut, the animals will munch them up too, and I understand that they are full of minerals.

The proof of the pudding is in the cattle themselves. They are plump and glossy-coated, meandering up to say hello, but calm and content. They are always curious and rather wary of Dog, so we stay in the next-door field, and they follow us along the fence line. Skylarks erupt periodically in front of us, soaring up into the evening air.

I come down from the ridge, past the ewes and lambs, along the rough track past the ancient shepherd's hut. The spring crops are up and away now on Hedge End and Lower Whitehill, though still weeks behind where they should be. It will be interesting to see how much ground they can make up, and what sort of yield we get at the end of the day. Then I suddenly see a pair of ears: a fallow doe lying under the hedge. Deer chasing was always one of Dog's favourite pastimes, second only to cat chasing; not one I encouraged, obviously. I am sure he won't have the energy for this now, but to be safe, I slip his lead on – there could be a fawn around too. And it was just as well I did, for as the doe sprints, the years roll away from old Dog, and he strains at the leash.

*

By Friday lunchtime, I'm wondering what all the stress and worry has been about. We have had a very sensible meeting with our landlords, and while we haven't settled it all, we have a clear way forward and an agreement to meet in a couple of weeks to continue the discussions. I just wish we had been able to have this conversation three months ago, to reassure us that our landlords are supportive of our plans. As with most things in life, communication is key. In the long-term partnership between us, one that has already lasted 68 years, and hopefully will continue through

Sophie's lifetime at least, we need to build plans together, and ensure that both parties are getting most of what they need from the relationship. This will be even more important as the whole landscape of farming changes after we leave the EU.

To wind down, I go to check on the orphans, and make a surprising discovery. The piglets who are in 'race training' have joined them, and seem to have made themselves entirely at home, languishing in the shade around the water trough, and trotting back and forth to the allotment. David has obviously read my mind and decided that this is a better spot for them than the barn by the offices, and I'm quite pleased. It is about time the orphans were integrated back into pig society, and it does seem that this has been a success. Our orphan pigs are much bigger than most of the racers, though there is also a larger, rather smart Saddleback boar in with them too, who I assume has been separated out in preparation for his future as a stud. The orphans tolerate the newcomers, gently curious and somewhat lofty in their attitude.

There is plenty of space for them all, and I hope that our friendly duo will help the motley crew get used to us humans. Right now they are not too keen, and certainly haven't got the banana habit yet. Which is probably just as well, as our grocery bills are going through the roof.

The pigs have seen more of Tim this week than I have. We have Belinda Carlisle singing at the pub tonight in aid of an animal welfare charity, and along with the normal hustle and bustle of this busy place, there's a military operation of planning and security to worry about. As silage trailers roar around the outskirts of the village and the sound checks reverberate from the garden of the Royal Oak, I am aware that the energy and life that the farm and pub bring to the village may be seen as disruption by some of our community. It's a glorious evening, though. Warm sunshine, a relaxed vibe, despite the bouncers on every entrance, and the pub garden full of people eating the vegetarian feast we have prepared (no meat was the order of the day from the charity, which disappoints some of the guests who know we are famed for the most delicious pork in the world!) while drinking Arkell's Ales and organic wines.

There's no relaxation on the farm, though. The team work through the weekend to get the silage in; we want to make the most of this glorious weather, and the grass is drying out quickly after cutting. There's an optimum moisture content for ensiling grass; too wet and you are carting water around, and potentially polluting effluent will run from the clamp for weeks; too dry and it's hard to get the consolidation you need to exclude the air and get a good fermentation. On Tuesday, the job is done. There are 151 loads of grass in the clamp, not as many as usual for a first cut, but we have started earlier than we normally do, and the grass itself is drier than it should be too. The black plastic sheet is rolled down one last time, and secured with old tyres and sandbags – a laborious, whole-team job for the tired crew. By way of small celebration, Henry brings out a cooler of beer bottles.

Our decision to grab the week of great weather is vindicated by Thursday, when it starts to rain. This is brilliant timing, as the ground is now bone dry, and the grass growth has slowed. We have spread cattle slurry on the pastures immediately after cutting, and the rain washes in the manure; with warm, wet conditions, we should be on track for a good second cut in six weeks' time.

Farming is such a cycle. All summer we are preparing for the winter, conserving grass, clearing sheds of manures, baling thousands of tonnes of straw. Every month has its task list, but with the constant uncertainty of the weather trumping all our planning. Timing is everything, grabbing the opportunities we have, being ready for them and making the most of them. Patience too; to wait until the moment is right, and having faith that the conditions will come right, even if not to our command!

The pigs are also pleased to have some rain. Their paddocks are dry, and the ground too hard to root in, except around the water tanks and wallows. I noticed Molly walking slightly stiffly across the cement-like ground around her arc last weekend, but today she is moving freely again, so I am reassured that there's nothing amiss. It is her five-month birthday, and I cut a load of fresh grass from the strip around her paddock to make up for the absence of cake. She and her comrades munch happily on the mounds I create along the fence line. They have food on tap, and silage too, but they seem thrilled by a change from their staples.

The largest in the pen must be over 90 kg now, and yet they are all still as friendly as they were before weaning, in those harsh days when the snowstorms raged and we huddled together in the arcs. Their friendliness is hazardous now. They crowd around me, rubbing against my legs, vying for attention and boot laces. I am in constant danger of falling over, so retreat to the boar pen. The boars are politely interested, and equally grateful for fresh grass, but never display the rather overwhelming enthusiasm that the girls do.

We need to keep moving with our ideas for next winter, how to stop churning up the land so much while keeping the pigs stress-free and content. I'm keen to involve the pig team in thinking this through, so on Friday we gather together to both review recent performance and so I can pick their brains on the future. We debate several options. We could try to build a shed with an outdoor run – a bit like the Forty buildings, but much larger. This would be very expensive, and I cannot think where on the farm we would get planning permission. And even with the best design, we know that the pigs are happier when they have soil to root in. David suggests a large tent in the middle of a field, where the pigs could feed and lie under cover – and where all the feed and straw could be stored – but still be able to range into paddocks that would radiate from this central point. It would need to be a very big tent, though, and would anything of this size stand up to storms?

We could move the growing pigs to lighter land somewhere, but where? Perhaps on the new land that we are planning to take on, but how much would it complicate things to have the herd in two places? Or we could try to adapt our current system, keeping the growing pigs fully outside, but prepare all the winter paddocks during the autumn, putting all the straw and possibly much of the rest of the equipment out during the autumn, so reducing the need for tractors as much as possible. It's the tractors, not the pigs, that do the damage.

The sows and their piglets consume a relatively small amount of feed compared to the weaned, growing pigs who eat tonnes of the stuff. We could use a much lighter vehicle, a Mule or mini tractor with a half-tonne feed hopper to feed them each day, and again, we could put all the straw in the paddocks before the piglets are born. If we then had two big feed hoppers in the weaned-pigs' paddocks, we would only need to replenish them once every ten days or so, and therefore could walk round each day to do the checks. If we got really organized, we would hardly ever have to take tractors on to the fields, except to move pigs, and fill the hoppers. It would mean buying more equipment, and I'm sure we would still need to move some around at times, but at least we could pick our days – when the ground is frozen, or during dry spells. It would involve more manual work, bedding-up by hand, and walking rather than driving from pen to pen, but the team seem up for that. What they would need, though, is a hut in the field to take breaks in, maybe an old caravan. A tractor cabin is protection from the elements, the only warm haven in our open country, and it becomes clear that it's the loss of that protection which worries them most. I completely understand, having spent much more time than usual with my pigs over this harsh winter. It should be an easy thing to organize, compared with the other options, and Dai undertakes to spend some time calculating the equipment we will need and planning how this could work in more detail. I sense we all feel it's the best solution for the farm and the pigs, though it would involve using as much, or even more, land each year, and that is going to be a scarcer resource with the establishment of the new dairy.

It's been a very busy and productive couple of weeks. Everyone is tired, but also excited by the progress we are making both in getting the work done, and planning for the future. The pieces of the jigsaw are starting to fall into place, in theory at least. There's still much to do in practice. I take the short walk from the offices to the pig mansion to see how the racing pigs are coming along. This evening they charge over to see me, already trained to know that people mean food, and much less wary in their outlook than they were even a few days ago. The orphans tower over them and demand special treats. They have now made numerous paths through the overgrown grass, and they show me their trails, all of which serve to connect the allotment with the water trough. We potter back and forth for a bit in the light drizzle, and as the rain gets harder, hide in the turkey house. Everyone settles down in the straw, mostly in a circle around me, while the mini-monsoon drums on the tin roof. My phones beeps as a text comes through. It's from Sophie: 'Passed all my exams! Xxx'. I call her, and sense, though she doesn't say, that she has done much better than 'pass'. An hour later we are celebrating this and much more too, with the many friends and locals who make Friday nights special in our overflowing pub.

20

A Day at the Races

Sunday 3 June dawns grey and cloudy. Perfect racing conditions, if not so good for beer drinking. Underfoot, the tarmac is good to firm, although light rain forecast for later might make it a bit soft and skiddy. By 8 a.m., the bookmaker, Tom Green, is at work erecting his tiny tent, fastening Piggy Power posters everywhere suitable. Clive, pig trainer and track builder, is already blocking the road and lining the track with his special sheep hurdles, designed to prevent any pigs veering off-course into the neighbours' gardens, where one minute's unfettered rooting can wreck all that spring planting. Sensible neighbours who have no interest in pig racing have heeded our advice and made themselves scarce, finding families elsewhere who mysteriously want to see them for lunch. On the whole, though, our neighbours appear to enjoy our daft days, and most of them are willing participants. Our somewhat crazy events at the pub are deemed by most to be a joyful expression of Bishopstone life!

We took over the running of the Royal Oak 11 years ago, mainly because it was run-down and in danger of closing, like so many rural pubs. It seems essential for a village to have a thriving centre, a place for eating and drinking and gossiping, and we rather innocently thought we could revive this flagging venture between all the other jobs we had to do. Madness! Like farming, running a hostelry is a labour of love, with uncertain rewards. And the traumas of finding the right chefs and front-of-house staff, of balancing the desires of a very varied community – dart boards or a quiet space for coffee and reading is just one small example of many early quandaries – while keeping the finances straight, have at times driven us to distraction.

The Oak is owned by the Arkell family, a local family of brewers with around 100 pubs. Their support and confidence that we could make something special from this run-down place has been crucial, and last year we took another bold move together, expanding the dining area, and developing 12 fabulous bedrooms in a derelict building next door. These rooms are all themed on farm fields, with wallpapers, pictures and features that resonate from places that are special to us. So now the pub is humming seven days a week, from breakfast to late at night, from walkers exploring the Ridgeway, to parties of friends meeting up for celebrations and reunions, to business people working in Swindon. Thank goodness, we have had wonderful reviews in all the national papers, and now we are developing the farm to allow people to fully and safely enjoy it while they are here. There are fabulous places to walk, run, ride, cycle and picnic. Or they can just curl up and relax in the Wallow with its revamped record player and comfy sofas, perhaps next to a blazing log fire in winter, in the pub itself. With our new photography expert joining the team shortly, the hides on the farm are getting a makeover, in the hope that visitors will enjoy the chance to observe and record the amazing wildlife that our farming approach encourages to thrive here. And, of course, most popular of all are the pig safaris. Tim is out most mornings taking groups around the farm in our ancient red Land Rover, regaling all with tales from this busy and occasionally chaotic place.

The village community has mostly embraced this hive of activity, and as farmers, we feel at the heart of life here as a result of taking on this crazy venture. We have planted a wood nearby where people can safely let their dogs run free, and have got to know even the most recent settlers here as a result of being the publicans. Or at least Tim has. This is very much his baby, and he throws his heart and soul into ensuring that our guests have a fabulous time, eating food from the farm or local organic producers, making the most of this rather undiscovered part of the country and taking part in the stream of events that litter our annual calendar.

This particular day of madness, now known globally as the International Pig Racing Festival, all started in 2012, when we received a note from the local council, via the great and good on the village parish council, advising us that since it was the Queen's Golden Jubilee at the end of May, we were effectively being granted a licence to do something, anything, that might assist in our village's obvious wish to celebrate the day. They'd even give us permission to close a road or two, if we wanted.

Suggestions from the council ranged from fetes to street parties, neither of which seemed very original, and if there's one thing Mr Finney likes to be, it's original.

Without even a word to me, Tim put a small notice on the blackboard in the pub: 'Pig Racing, May 31: The Queen's Golden Jubilee Handicap'. Nothing more than that, just to test the waters. I failed to spot this for a week or two, and when I did, I was not convinced, to say the least. I imagined scared pigs running amok, complaints from residents, health and safety issues for people and pigs. 'Too late,' he told me. He'd been knocked over in the rush to help him establish this event, with customers, staff and neighbours all wanting to know what they could do. Four local charities had also offered to be involved.

My nervousness sobered him slightly, but did not deflect him from his self-appointed task, this highly unlikely Royalist. He advised that six pigs, carefully selected on the grounds simply of recognizably different colours (he was doing the race commentary of course) would be racing the length of Cues Lane outside the pub, just once, and to make it more fun, there'd be some betting, all in aid of the local charities, including our wonderful local hospice, and the Injured Jockey's Fund.

The rest, as they say, is history; well, Bishopstone history at any rate. As Tim puts it in his now annual pre-event marketing, the International Pig Racing Festival has become part of the summer sporting calendar for people who have to be seen in all the right places, slotting neatly between the French Open, Royal Ascot, Henley and Wimbledon, and usually just before the other big race at Silverstone. If you listen to him, he'd have you believe that the invisible powers that select and agree dates for their international sporting events years in advance now consult him to make sure that their event is not going to see its attendance halved because of the clash with pig racing.

Over seven years, we've had a total of some 4,000 people head for the natural sporting amphitheatre that Cues lane, Bishopstone, and the Royal Oak gardens, most definitely are not, but that is where they gather. Never has a location been more unsuitable, and yet more popular, for a day of pointless entertainment, classic English style. The lane itself is narrow, with steep and slippery grassy banks, and officially closed by order of the Queen from 10 a.m. till 4 p.m. It is slightly curved so no one can see the whole length of the 127-metre racetrack, not even the commentator. He has to ask the crowd what's going on at times, as he can't see the finish line, and if he could, he wouldn't recognize which pig had won.

'And the winner is . . . sorry, what was that . . . ? The one with the brown spots behind the ear . . . ? Hold on a minute . . . it's Albert Einswine . . . Albert Einswine takes the seventh Queen's Jubilee Handicap . . .' At which point part of the massive Bishopstone throng yelps with joy and surges towards the Piggy Power betting tent to demand their winnings, often quite impatiently. For every five-year-old that won ten pounds on Albert Einswine, there are 20 small children bemoaning their loss of 50p. Thus are early betting habits crushed. Children under ten have to be accompanied by an adult when they place their bets, of course.

The queue for beer snakes out of the pub front door on to

the racetrack, which starts, as you'd expect, by the telephone box under the hedge. The finish line, where we have official photographers on standby to capture the astonishingly close races, is a red line painted in the road just below the VIP lawn. The VIP lawn is a patch of grass in front of the neighbours' house, bang next to the pub terrace. Tom and Millie Green and their daughter Tabby run the all-day betting stall from here, as well as a cake stall, naturally, and allow select guests to stand behind a piece of string, from where they can see just a part of the action.

The finishing line itself is a much more important part of the day than you'd probably imagine. One of the oddities of pigs, despite, or because of, their intelligence, is their discomfort at any apparent change in the surface under their feet. As I've already mentioned, whenever we move our pigs – into trailers, lorries and the like – we have to disguise the floor beneath them, usually with straw. When they've been used to grass and soft earth under their feet, they are suspicious of a metal ramp, or even a wooden surface. They stop, have a look and sniff, and after a while, one of them steps over or on to it, and the rest follow, assuming the first one doesn't change its mind. Hence the entertainment of a red line, not that Tim had thought about this before he painted it.

It's not a change of surface of course, just a different colour on the tarmac – but to them, it's different in some way. Hence, even in a race where all the pigs are well strung out, with some of them still taking a drink from the water jump and another one having a pee en route, the first pig arrives at the red line and stops, rather like something out of the Woody Woodpecker cartoons from years ago. The head might cross the line, but the feet skid to a standstill. And winning a pig race is all about the feet. So the rest of the pack follow at their own pace, see that the leading pig has stopped, and they all stop too. They all have a careful look and sniff at the red line, until one of them – it might be the one that stopped for a pee earlier – decides that it's not dangerous or risky, and steps over, and we have our winner!

Only once has the red line been ignored, in 28 races involving about 60 pigs over seven years. Possibly our most famous and successful racing pig, the Westland Flyer, unimaginatively-named after a local dog rescue centre, for some reason got the hang of pig racing very early on in 2016. After loitering midfield in the first race, he was clearly stung by a wasp in the second one, and charged towards the finish, winning by at least 50 yards. The red line was obviously never seen. The third and fourth races, possibly encouraged by the prize of one kilogram of organic pig nuts, Westland Flyer took by similar margins. There was an outcry from the punters.

The training of racing pigs is not an arduous task, and it's certainly not designed to make them go faster than they want to, or even in the right direction. One of the benefits of farming the way we do is the amount of human contact that our livestock enjoy. Not just from the pig farm staff, but from student visits, other farmers and office staff, and of course our pub and hotel guests and customers. In spring and summertime there are dozens of people each week having a nose around, so there is little to shock our pigs when their trailer arrives at the Festival and 100 people put their heads over the side, and some kids climb in.

The selected pigs have also had a few weeks of special handling, learning to come when a bucket is rattled, getting used to the attentions of a whole variety of people from the comfort of their luxury accommodation close to the offices, accommodation that is appropriate to their thoroughbred status. Which is just as well, because as they are released from the trailer, there is, by Bishopstone standards, a wall of noise from the hundreds of people lining Cues Lane behind the specially erected crash barriers. What the pigs are really looking for – as they open their eyes, since quite often they've been asleep in the trailer before being ushered out for the race – is the bucket with organic pig feed inside it. Food is a very important part of their lives, of course, and they'll go anywhere, in any direction, if there's something to eat when they get there.

Money is also raised through sponsorship – customers far and wide clamour for the opportunity to 'buy' a pig for £40. In return, they are given naming rights. If they can, they visit the racers and choose one of them for themselves. On top of this, Tim grants them the honour of being abused in the race programme notes for the day. Any story about the hapless family or individual owner, whether true or false, is blown out of all proportion for the purposes of harmless entertainment. My initial concerns have long been dispelled, and this ridiculous day has become one of my favourites. All the pigs have gone on to be well-rounded individuals, in every sense of the word, and most sponsors have taken the race cards, with the rude stories about them, home where they now sit framed in the downstairs toilet.

This year we have a special race, for our oversized orphans. They are too big for pig racing, really, but are so tame and so part of the family that it's inconceivable that they should be excluded from the fun and games. I have been trying to avoid admitting this, but Tim has coined them Jeff and Sharon for reasons that are impolitic to disclose, and this is how they appear in the race card. Alongside them, we have Gunter; One Pig Went to Mow; Oinkment; Albert Einswine; Perky; Princess Megham; The Carlinger and Cues Lane Flyer.

The first race is the usual shambles, with bemused pigs trying to fathom what new entertainment this is. They wander around the start line, looking up at the crowds, then Perky responds first to the sound of the feed rattling in Clive's bucket. He ambles, then trots in the right direction, and the others notice his enthusiasm and head off after him. Halfway down the course he becomes distracted by a beautiful girl with long blonde plaits leaning over the hurdle barriers, and stops for a chat. Cues Lane Flyer takes over the lead, with Gunter close at her heels – and it is these two that get to the red finishing line first. But, as is so often the case, they stop to check this treacherous line, and it is One Pig Went to Mow who finally pops his trotter boldly over to claim the prize.

A quick snack for all the pigs, then back to the trailer for 40 minutes to allow punters to garner their winnings, grab a drink, and lay their bets for the 3.48. This time Perky learns his lesson and is both first to make his move, and maintains his focus right to the line – and over it too.

The third race is the Bishopstone Hurdles, conducted over old broom heads. It is Princess Megham's turn to triumph, her longer legs proving a distinct advantage as she steps over the fences. As the roar dies away, the eight runners retire gracefully to the pig trailer, and Tim disappears. A moment later, there he is, trotting the wrong way up the track with his two orphan pigs in tow. They become instant starlets, loving the attention, turning left and right to greet their fans, posing for photos like divas on the red carpet before scampering after their master. There's only one person they are going to follow down the racetrack, so I take over the commentary.

From his voluminous brown striped jacket, Tim reveals a banana. Waving it aloft, I start the countdown, and away they go. It's over in a flash, which is just as well with my commentary skills, with the gilt – do I really have to call her Sharon? – beating her brother by a good yard. They share the banana, though, and the crowd are baying for more. Back they go, and a team of helpers reinstate the hurdles: now, this will be interesting. Another banana emerges, again we count down, this time the entire village in unison, and away they meander to the first fence. Both stop and peer over, or rather, look down over. 'C'mon, little piggies!' cries Tim, in a way that itself has the crowd in stitches, and with still no forward movement, starts to unpeel the banana. That does it, and with an ungainly hop, 'Jeff' allows greed to overcome suspicion, clears the fence and makes for the next barrier. Always reluctant to be separated, the gilt follows suit. By the third hurdle, they have quite got the hang of it, and the battle of the sexes is level pegging at the end of the day.

The afternoon continues with music and feasting, and the

two pigs do a tour of the gardens, taking plaudits from all and sundry, before being lured by yet another banana back to their paddock. Quite enough excitement for one day, for both the pigs and the village.

21

Summer Solstice

It is the longest day. A heat haze hovers over the pig field, and all is still. The only sound is a distant clatter from the hay tedder as it spins out the grass to dry. The acres of velvet barley are just starting to yellow; harvest is only a month away.

The sows lie heavy in their wallow, bellies swollen with growing piglets. Messy Bed, as ever, has the prime spot, spreadeagled in the centre like some grand duchess. Molly's mum, always more timid, is pushed out to one side, panting quietly in the heat. Only Mrs Duroc comes to greet me, covered in sticky mud. She's made such an effort that a good scratching is only polite, even though it leaves me almost as grubby as she is.

Green shades dot the field in each of the serving paddocks. It's vital to keep the boars' balls cool to preserve their fertility, so they and their sows are stretched out beneath the gauze sheets, which are held aloft by four angled poles. Only the young piglets seem impervious to the high temperatures, scampering around in long grass, which almost shades them, through a maze of trails across the paddocks.

The tracks between the paddocks are easy walking now, smooth and baked hard by the sun. Despite this, it feels hard-going. I am struck by the same lethargy as the pigs, hot and lazy on this humid summer's day. I meander through paddock after paddock, crossing the low fences, kicking soil away where it is shorting out the electric pulse. The growing pigs can hardly be bothered to do their mock-horror thing as I pass through, stunned by the heat and not wanting to give up the cool spots they have secured in wallows or in the shade of the arcs. Away to my right is a bare, dry, empty patch that pulls at me like a song that I cannot get out of my head. Molly's paddock.

The last two weeks have passed in a flash. We have had another meeting with our landlords, and are a step closer to agreeing what involvement they will have with the Starveall dairy. They are very keen to resuscitate the semi-derelict cottage in which everyone who sees it wants to live, in summer at least, and we have undertaken to help make this happen. We have paid the electricity company; they wouldn't give us a date to do the work until we paid them in full – if only all our businesses could run on similar lines. And next month the Kiwi builders arrive to start the dairy build, so we have contractors onsite, demolishing the remnants of the old sheds, levelling the site and preparing a cottage for the workforce to live in for the six weeks it should take them to build the parlour.

It's a huge project for us, and we have been worried, in the moments available to us for such luxuries, that despite our careful planning, there are myriad things that could go wrong. The builders have emphasized that any downtime they have that can be attributed to our incompetence will be billed at £1,500/day; we can't afford too many of those! But help is at hand, in the form of Sian, who joined us last week. We have been on a recruitment drive in the marketing business. After a couple of years of consistent growth and profitability, we have decided to invest in both sales and new-product development – including those dairy products that we have talked about for years, but not managed to do anything about. This feels like the time to be bold. Two years ago, we recruited a remarkable woman, Vicky McNicholas, to take on the leadership of our marketing business from Tim. Her calm and undemonstrative demeanour belies a sharp commercial brain, an ability to juggle a dozen balls, and deliver what you've asked for before you've finished the sentence. She is wise too, able to get the best out of her growing team, as well as mediate and mentor the rest of us when family dynamics are at risk of getting in the way of sensible business thinking. With her in charge, I have every confidence that this is the time to push on and grow. Yes, we have tried and failed a few times before over the decades, but now we have the best team ever, so maybe this time we will succeed.

*

The six Montbeliardes have finally been introduced to the 140 heifers who have been making eyes at them for weeks. The Irish girls have grown fast since they arrived, though our home-grown heifers, who have now been amalgamated with them, are still stronger cattle. But while a couple are still a bit small, if we can keep them on good clover pastures for the next few months they should continue to make up lost ground. The bulls are working hard, and tomorrow we will swap them with the Anguses, to give them a break for a week or so. If all goes to plan, the heifers will start calving in the second week of March.

The silage fields are mostly knee-deep in grass again, and we are starting to monitor the weather forecasts even more avidly than usual to determine when we should take the next cut. The spring-sown cereal crops are still rather paltry, though; they were planted so late that I fear they will never come to much. Meanwhile, this hot spell is the perfect moment to get on with hay making. We need plenty of soft meadow hay for our baby calves, and some more fibrous material to keep the cows' rumens working well during their annual dry periods, when they take a break from milking before they calve again, and to supplement the moist, acidic silage during the winter. Buzzards and red kites quarter above the hay fields, just as they do when we are making silage, awaiting any prey exposed by the activity below.

It's good to be here. Over the last few weeks I have been away a fair bit, in Italy to meet with other progressive organic organizations, in Scotland too, and, as always, back and forth to London and Bristol. After the pressure of the government's consultation on the future of farming, there has been a sense of limbo while we await the proposals for the Agriculture Bill, and try to get a better idea of the framework we will be working in. There is still no sense of what our trading relationship with the EU will be after we exit, or whether people will be able to come here to work, or any detail of any support that farmers may get to deliver environmental benefits and the other 'public goods' that will need to be funded by the taxpayer. We don't even know what the rules and regulations will be. In the face of all this uncertainty, my commitment to making our farming businesses as independent and self-sustaining as possible becomes ever stronger. We must continue to forge paths that allow us to connect directly with people who value what we are trying to do here, and want to support us, either by buying our products, or coming to eat and stay with us on the farm.

Likewise, the Soil Association needs an army of supporters to speak up for humane farming that cares for our health and for nature. Our founder talked about food and farming being at the forefront of our national health service – rather than our national disease service, which is the way things are largely organized now – and this feels like the moment when that philosophy could be put into practice much more widely. In this 'unfrozen moment' anything is possible, good or bad, and it's the choices we make over the coming months and years that will swing the balance. We are beginning a membership recruitment drive, to try to muster a bigger voice for an ethical, diverse and vibrant future for farming and food; it's a big investment for us, but as with the farm, this must be the moment to be bold. So much is at stake.

In the early evening I walk up through the Valley to catch the end of the orchids. They pepper the banks in great profusion, nestling alongside the quaking grass, ladies bedstraw, milkwort and kidney vetch. We moved the yearling cattle out a few weeks ago to allow the vegetation to recover and mature before we start to graze it with the dry cows in August. This lovely place, with its hundreds of species – not all of which I know the names of – seems to do wonders for our pregnant cows, who can select the plants they want and need from the smorgasbord underneath their feet here. This natural pasture will never yield the volumes of feed that our sown leys will, but it has great value nevertheless, and animals are always contented on it, even when there seems to be little to eat.

Swallows zip along the valley bottom, seeking the insects that proliferate, including those that are avidly breaking down the dung pats left by the cattle. I find my favourite spot and lie against the bank, watching planes overhead and feeling part of the earth.

The wind is picking up, and the early evening has an almost eerie feel to it. The light is yellowing, and the humid heat seems to condense into a threat of thunder. I move on up to the Ridgeway, and can see the hay-making team beavering away, stacking the bales they have made with some urgency, some on to trailers, the rest in blocks on the ground. The last thing we want is hay getting wet. Above them, the fields where we lambed the ewes in that appalling weather are now empty, regrowing a profusion of clovers, chicory and sainfoin. The ewes and their lambs have moved to the new land near Lambourn, to the rolling grasslands that will be their primary home from now on. From next spring, our young beef calves will join them, leaving more land here for the Starveall dairy herd, which will be grazing these downland pastures in future. I can't wait to see the cows here – a long-term dream at last a reality.

There is a palpable electric tension in the air, and I begin my march back down, away from the exposed open ground. I have left Dog at home – he has good days, when it's cool and he feels able to enjoy a decent walk, but in this hot, humid weather he is happier in the shade of the garden – so I can cut across fields and fences, for a while following the line marked out for the trench that will carry power to Starveall from the Ridgeway. It cuts straight through the pig paddocks, so there will be some disruption to their lives shortly.

By the time I reach the village, there are flashes of lightning in the distance, and the heat haze has thickened into grey overhead. I haven't seen the orphans today, so I take a shortcut through the pub car park on my way to their home, and find Tim in the garden there taking a few minutes away from the hubbub inside. We have hardly seen each other this week, so we saunter down together to catch up with our charges.

Instead of calling them when we arrive, we follow their trail to the allotment. Despite their big ears, it's remarkably easy to creep up on them without them hearing, and we watch them for a while as they go about their lives amongst the now well-rooted vegetable beds and redcurrant bushes. Most of the racing pigs have rejoined the herd, but the two Saddlebacks who are destined for breeding, rather than sausages, are still with them. The four of them have become firm friends, and the sight of them mooching about reminds me of Beatrix Potter, the naughty rabbits in Mr MacGregor's garden. When we do call out, they startle in almost guilty fashion, and then hurtle across to greet us. Or rather, to greet Tim. He is clearly the favoured one, after his constant attentions over the last few months, and the crates of eggs and bananas.

They follow us back to the turkey house, and we sit on the

rickety chairs with the four of them vying for affection. The gilt just loves it when Tim runs his fingers down her spine, and she starts to slide earthwards, until her belly is entirely exposed. He rubs her tummy and she squirms with pleasure. I'm impressed by what she has shown us, a strong line of 16 teats. We always ensure that gilts for breeding have at least 14; with this number, and her maternal genes, she should be a brilliant breeding sow. Now we just need to decide what to do with her brother. If he is going to continue to live with his sister, he should be either castrated or vasectomized fairly soon. His hormones are starting to rage. At times he is as affectionate as his sister, but at others he is aloof: a sulky teenager. Just as I have seen in the pig field, as they get older the boar pigs have very different characteristics compared with the gilts. Less affectionate, with each other and with humans, and more competitive too.

From behind the hut, Tim produces a rather strange-looking harness, and Number 3 pig seems entirely happy to have it fitted on to her. They trot off around the paddock together in perfect step; this is clearly the result of several weeks' secret training. Again, I'm impressed. I can see that it won't be long until the two of them are parading the village together, and I'm rather glad that Dog is at home. He would be deeply offended.

Tim needs to get back to our busy pub, but I stay for a little longer, sitting inside the arc as a few heavy raindrops sputter from the darkening sky. The storm is getting closer, but it is still warm, the air moist enough to drink. I can hear the clang of the feeder lids from the Forty buildings just 100 yards away. I want, and I don't want, to go over there before the skies open.

The building is quite full, with Molly and her female friends running between two pens, connected by the outdoor run. In the other two pens are the boars. They are on deep straw, though the area nearest the outside is mucky and wet, and quite a few of the pigs are lying and rooting in their makeshift wallow. The rest are mostly feeding, making up for their overnight fast yesterday. The only way to catch pigs without stress is to lure them with feed, so Tony will have taken the feeders out of their paddocks to ensure that they are hungry enough to follow him into the trailer. 'Feed is our best friend' is one of David's favourite sayings.

There's still quite a disparity in size in these groups. Tess's offspring are much smaller, though looking healthy, and they will be split out from the group in the morning, and returned to the field to grow on for a few more weeks. But this tale can only have one ending for Molly and her gang, and they will leave for the abattoir tomorrow. I sit on a mound of straw in the corner of her pen, and for the last time they crowd around me, enjoying a scratch, chewing my boots, until the majority get bored and wander off. The thunder is getting closer, and flashes of lightning illuminate the darkening sky. The rain becomes heavier, and water runs down the track from the dairy, washing away the hardened grime, reviving the parched pastures, renewing life on this patch of the earth. Then the floodgates open, and the world is all water.

Epilogue

The last six months have been as much of an adventure for me as they have been for the piglets and their mothers. Although I have kept pigs for over 30 years, the excuse to observe one group so intimately, to spend time with them almost daily, has never arisen before. I have spent so much of my life trying to find practical ways of improving the lives of farm animals, especially pigs, and campaigning for this too – but this has been for pigs in general, because it's the right thing to do. The experience of getting to know *specific* pigs, of attaching to them so that at times I almost felt like part of their community, of being with them through all their trials and tribulations, has allowed me to both question and ultimately to validate my commitment to an organic way of keeping these wonderful animals while inspiring me as to how to improve their lives further.

As the pig team have remarked on several occasions, I could not have chosen a more complex group with which to spend my time. We do not usually have fox attacks, sow death, blizzards and piglet abandonment all going on in the same gang, I'm pleased to say! The hazards that this group faced were extraordinary, and living with them through these traumas forced me to face the downsides of keeping pigs in more natural conditions. As I try to draw out in this book, there are strong parallels with the way we view risk in society, and especially in the way we rear our own children. Do we confine and constrain them, or allow them to climb the apple tree despite the danger of falling out? At the end of the day, I feel it has to be a balance of avoiding crazy risks, but accepting that for us to be fully alive, to become fully human, we must explore, test ourselves against the elements, experience reality – and for me, that reality is found in the natural world. Humanity's divorce from that real world, our retreat into a romanticized perspective of nature, and a fear of its elemental power, can leave us perplexed.

At least as humans we can make these choices, though this is harder for some than others. The same is not true for our farm animals. We determine the lives that they will lead, whether they will have the opportunity to experience the highs and lows of an outdoor life, or whether they will be simply units of production in a barren concrete environment, with no chance to express their innate behaviours. Some people justify the ultra-intensive farming of pigs and other species on climate-change grounds, in that animals will use less feed if prevented from ranging freely. My take on this is that if we cannot allow our animals a good life, we should not be eating them at all. Humans can thrive without meat, and there are ethical red lines we should not cross in the way we treat other species.

It may seem strange to some that I have loved and then slaughtered my pigs. It is hard, and so it should be; but as we see in the cave paintings from ancient times and even in certain hunter-gatherer societies today, reverence and utility can exist in the same breath or brushstroke. A hunter honours his prey. Being fully conscious of what we do, doing it as well as we possibly can, respecting the age-old contract with the animals we domesticate: these things are important to me. I am still happy to eat meat, even from those animals that I have known so intimately and cared for, but only if the animal has lived well. I choose to eat less of it than I once did because of the environmental impacts. 'Less but better' is my mantra.

Farmers live close to the seasons, and the weather is our greatest uncertainty. Re-reading this diary, stretching from the winter to the summer solstice, from that dark, still, introverted moment at the turning of the year, to the almost endless light and energy of midsummer, I wonder whether the urban reader may be bemused by the constant focus on weather. Yet every day, it determines what we can do on the farm, and how we will do it, and it also underscores the valiance of the heroes of this tale, the people who work uncomplainingly to care for the pigs, whatever the conditions. Our farm workers, at Eastbrook and across the globe, deserve such respect and admiration.

We are sometimes lauded as a success story, but I always demur. We have persisted and survived, and that in farming is success, while in most businesses one is judged by the size of your profits and capital assets. To be honest, we could have done better on that count. We have made mistakes, many of them, and often been too ahead of the game with our ideas, before the market was ready, or maybe not delivering the ideas as well as we could have done. We have chosen a difficult business model, and we have tried not to compromise on the things we most care about, especially the welfare of our pigs. There are many upsides of following your star, though, and one of them is the huge fun we have had along the way, and the many brilliant people we have worked with, both here and in the businesses we partner with. It's been a roller-coaster ride at times, and one that I would not have missed for the world.

So what have I learned about pigs? The thing that has struck me most is just how competitive they are with each other – deeply selfish when the chips are down, or indeed, any other form of more appropriate feed. They have an innate drive to eat and grow, and very little, including each other, will stand in their way. How different the boars and gilts are, especially as the hormones start to rage; the gilt pigs remain very sociable and friendly, while the boars clearly have other things on their minds, and lose interest in humans almost entirely in their teenage months. But once this rampant period is over, both the boars and sows develop less generic traits, an individuality that is not so apparent in their early life, or is at least masked by their drive to survive and breed during this part of their lives. Mrs Messy Bed is an especially strong example of just how characterful they can be! My time with them also reinforced just how much of their lives is experienced through their noses, those ultra-strong, sensitive digging machines; any system of farming pigs that does not allow them to root in a suitable material, ideally soil, can never be described as high welfare.

This story spans the life of the pigs, and ends when they do. But of course the parallel tales of the changes taking place on the farm, with Sophie and Dai starting to determine much of its future, and the far-reaching implications of the debates about the future of UK food and farming that have been triggered by our decision to leave the EU, are far from concluded. In both cases, these are age-old debates. The passing of the baton to the next generation, the mix of pride and trepidation, are universal experiences often fraught with tension – and at the very least highlight the poignancy of recognizing the turning of life's wheel. No doubt this story has many more twists and turns ahead for us as we seek to combine what we love to do with what makes financial sense to do, while anticipating future consumer trends and the verdicts of our political masters. Like the pigs, who are programmed to fight for survival against all the odds, family businesses like ours must be similarly inventive and tenacious. Like the piglets, I hope we have our survival tactics right!

Meanwhile, NGOs like the Soil Association will continue to make the case, and show what can be achieved on our farms, in our schools and hospitals, our forests and fields, if given half a chance. Our task is not just to show a better way of farming, but to make that better way even better. My exploration with trees is a small part of that endeavour. While it seems so simple, when seen alongside and allied to the care of our soils, the widespread planting of trees for fruit, nuts, biomass and timber could be the basis of a revolution for climate change, our diets, water management and wildlife. What's this got to do with the pigs, I hear you ask? Well, of the very many things that was hammered home to me by my six months in the pig field, one was how much time we spent trying to protect them from the wind – blowing snow and rain into arcs – or the sun. And how much the pigs loved to escape to root and play in the hedges at any opportunity. While our open-field, nomadic pig system has many merits, it would be much enhanced by more trees, to cut the wind and provide natural shade, even feed. If we could achieve only one thing in the turbulent years ahead, knocking down the barriers to tree planting and securing support for organic farming combined with agro-forestry and other ecological methods across the country, then I would feel more confident of the future for us, for wildlife and for a free-ranging, good life for our farm animals too.

And while the political debates seem very current, that's not really the case. Many of them stretch back over centuries, playing out in all parts of the globe: the extent to which we should protect our industries, including farming, and strive for increasing national self-sufficiency; how we get the balance right between consumer affordability, fair trade, and the safety and ethical standards we should abide by; the assumed tensions between environmental and social objectives.

I had a recurrent dream as a child. My house was burning, and I was running to get there but was constantly distracted by things along the way, so that I would forget the emergency for a while, and then remember, and restart my frantic journey. I have no idea what I was going to do about the burning house, or whatever it was that I had to do, but that's not really the point. For me, this dream felt like a premonition of my life, how easy it is to lose track of what you set out to do.

This is the challenge for leadership: it must look beyond the political cycle, or the next quarter's financial reporting, and invest long-term in preventing the problems rather than being distracted by short-term fire-fighting. It must put *values* rather than just 'value' at the heart of our food system. Can we, will we, grasp this opportunity to set a better course for humanity, one that accepts our interdependence with the natural world, and our responsibility to use our power ethically? Now, more than ever, I sense that the world is hungry for this visionary but practical leadership. But leadership is a shared responsibility. The onus is on all of us – politicians, pioneers, farmers, consumers, activists in our own spheres – to provide it.

Acknowledgements

Many thanks to all those who have been supportive through the last few months, as I have diverted every spare moment into both observing and writing about my pigs. Some duties have been rather neglected, and friends have had short shift. The encouragement from those who read early chapters and didn't tell me to abandon the project was hugely helpful. This includes my sisters, Pip and Lucy, my daughter Sophie and Tim's talented son Sam, who also took so many of the photos that you will see on our website. Thanks also to Jeanette Orrey, Jon Alexander and Peter Melchett, both for the amazing work they do, and for their support in this and many other projects over the years.

Thanks are also due to the farm pig team, David Henderson, Tony Connolly and Rachel Neaves, who graciously put up with my constant presence in the field but, more importantly, got the herd through such a grim winter. Their care and dedication deserves medals.

Without the initial persuasion of Alex Clarke and Auriol Bishop at Headline, *Pig* might never have existed. Alex, and his colleague Shoaib Rokadiya, worked hard to get this book over the line in the ridiculously tight timescales we gave ourselves, and Celine Kelly's wonderful advice to this novice author has been invaluable. It's been a great team effort.

Finally, to Tim, my co-author and partner in crime. He wrote reams of brilliant stuff, especially on the trials and tribulations of our attempts to make and sell products, and also on the craziness of our current food system, only fragments of which appear here. His dedication to the orphans, and his encouragement through the whole process, has been extraordinary. This book would never have seen the light of day without him.

About the Authors

Helen Browning runs Eastbrook Farm in Wiltshire, home to her beloved pig herd and many other animals, as well as the Helen Browning's Organic brand. Since 2011 she has also been Chief Executive of the Soil Association, a charity which campaigns for a more humane, healthy and organic future for farming and food.

Tim Finney grew up in Bradford and studied Agricultural Economics at Aberdeen University. He was an editor and presenter for BBC Radio 4's farming and environment programmes, before joining Helen Browning's businesses in 1995. Today, he runs the Royal Oak – Eastbrook Farm's dining pub – and Helen Browning's Chop House in Swindon. He lives with Helen Browning, much to his (and her) surprise.

Join the Soil Association and come to meet the pigs!

As well as looking after her beloved Saddlebacks and all the other animals at Eastbrook, Helen is also Chief Executive of the Soil Association, the charity dedicated to humane, healthy food and farming. Here she brings her organic expertise to a cause she is passionate about: ending the practice of farm animals being kept in cages, crates and concrete, and ensuring that all farm animals have as good a life as possible.

At the Soil Association, we believe every animal has the right to a good life: a more natural life where they can graze, forage, root and play, where they can feel the sun on their backs, and are only given antibiotics when they really need them. If you feel the same way, then please consider joining the Soil Association. The more members and supporters we have alongside us, the more we can do to give farm animals the chance of a stimulating life – sometimes with more risk and challenge, as you will have gathered from this story, but with so much more opportunity for fun and pleasure than living confined, bored and stressed in a concrete jungle.

With support from people like you, we've already improved the lives of hundreds of thousands of cows through developing and promoting welfare outcome assessment, an objective way of scoring important measures of animal wellbeing, like lameness. Our campaigning has put a stop to the preventative use of antibiotics in the poultry sector, protecting vital human medicines. And our work with farmers is improving the lives of animals through supporting their grassroots research.

But there's loads more to do, especially at this critical time as the future of farming in the UK is being decided. Issues like the trade regime we will be subject to and the kinds of support – if any – that farmers will get, will determine the future of our farmers, countryside and farm animals.

By joining the Soil Association and giving a regular donation each month, you will be adding your voice to our call for higher animal welfare standards. You will be telling farmers, retailers and the government that you believe all farm animals deserve a good life. And as a special 'thank you' to readers of this book, if you sign up today at the link below and are able to give £5 a month (or even more), you will be invited as my guest to visit Eastbrook and meet the pigs. Come soon, before the orphans start their careers as truffle hunters, or whatever their next adventures are to be, and before I lose patience with Mrs Messy Bed altogether!

If you share our vision of a better life for all farm animals, then join us today and together we will make a huge and lasting difference to the lives of millions of animals. And I look forward to meeting you at the farm, perhaps.

Thank you!

Join us here: www.soilassociation.org/pigbook

At the Soil Association, we care for animal welfare, our soil, wildlife, water, forests, and human health. By joining us, you'll help change the world's food and farming for the better.